PROGRESS IN COLLOID & POLYMER SCIENCE
Editors: H.-G. Kilian (Ulm) and G. Lagaly (Kiel)

Volume 85 (1991)

Polymeric Layers

Guest Editor: P. Wünsche (Leipzig)

Springer-Verlag
Berlin Heidelberg GmbH

ISBN 978-3-662-15684-1 ISBN 978-3-7985-1684-7 (eBook)
DOI 10.1007/978-3-7985-1684-7
ISSN 0340-255 X

Originally published by Dr. Dietrich Steinkopff Verlag GmbH & Co. KG, Darmstadt in 1991
Softcover reprint of the hardcover 1st edition 1991

Chemistry editor: Dr. Maria Magdalene Nabbe; English editor: James Willis; Production: Holger Frey.

Type-Setting: Graphische Texterfassung, Hans Vilhard, D-6126 Brombachtal

Preface

The investigation of layer properties especially of polymeric layers is a scientific field with a quickly expanding volume considering the strongly increasing number of publications.

So the 7th International Seminar on Polymer Physics was concerned both with the microscopic structure, the dynamic behaviour and the macroscopic properties of polymeric layers. The during the seminar discussed layers were not only ultrathin amphiphilic monolayers or Langmuir-Blodgett films of proteins but also macroscopic films with a thickness in the order of 100 µm. The majority of at the seminar presented lectures is summarized in this volume together with the most important discussions to review the up-to-date results of physicists and chemists working in the field of polymeric layers in many European universities and institutes.

This seminar was the first one held in the unified germany in Güntersberge, a small village in the Harz mountains. The familiar atmosphere stimulated interesting discussions after every lecture accompanied by additional discourses between the participants. So this seminar again was enabled interchanging scientific ideas and bringing together scientists of different countries.

Peter Wünsche (Leipzig)

Contents

Preface V

Chudáček I, Slavinská D: Electromechanical behavior on thermoplastic with modified surface (Frost effect) . 1

Paul E, Heise B, Schrodi W, Kilian HG: Deformation of semicrystalline and molten polyethylene the role of entanglements ... 12

Wübbenhorst M, Wünsche P: Inhomogeneous distributions of polarization in polyvinylidene fluoride mono- and multilayer films studied by the laser intensity modulation method ... 23

Birshtein TM, Borisov OV, Mercurieva AA, Zhulina EB: Thin monomolecular polymer layers as low-dimensional polymer systems ... 38

Brehmer L: Progress in Langmuir-Blodgett-electronics ... 46

Erokhin V, Feigin LA: Deposition and investigation of protein Langmuir-Blodgett films ... 47

Peterson IR, Möhwald H: The organization of aliphatic chains in ultra-thin layers and its importance for layer properties ... 52

Stamm M: Accurate measurement of the interface width and density profile between polymers by reflectivity techniques ... 55

Köpp E, Pechhold W, Sautter E: π/A-Isotherms of polymer thin films spread on a Langmuir trough and its discussion on a supramolecular level ... 59

Holstein P, Spěvaček J, Geschke D, Thiele V: Solid state NMR investigations of polyamide 11 films ... 60

Althausen D, Wünsche P, Peskova E: Dipole reorientation in polymeric layers investigated by photo-induced current ... 66

Heckner KH, Uhlig A: Mechanism of the formation of polyaniline films at transparent conducting electrodes and characteristics of their optical and electrochemical properties ... 75

Plümer F: Defect morphology in preoriented systems with constrained perturbations ... 76

Hergeth WD, Steinau UJ, Bittrich HJ, Schmutzler K, Wartewig S: Submicron particles with thin polymer shells 82

Wróbel AM, Kryszewski M: Preparation, structure, and some properties of organosilicon thin polymer films obtained by plasma polymerization ... 91

Heilmann A, Hamann C: Deposition, structure, and properties of plasma polymer metal composite films... 102

Weichart J, Müller J: Preparation and characterization of glassy plasma polymer membranes ... 111

Biedermann H: Basic physical properties of thin films prepared by unconventional techniques ... 118

Havránek A, Slavínská D: Various structure of plasma polymerized layers ... 119

Kremer F, Vallerien SU, Kapitza H, Poths H, Zentel R: Molecular dynamics in ferroelectric liquid crystals: From low molar to polymeric and elastomeric systems ... 124

Freidzon YS, Shibaev VP: Optical properties of LC polymeric films with helical structur ... 125

Wendorff JH: Optical and electro-optical properties of polymeric liquid crystals ... 126

Geschke D, Fleischer G: Orientational order in LCP layers ... 127

Gerhard-Multhaupt R: Thin polymeric layers for spatial light modulators ... 133

Schrader S, Koch KH, Mathy A, Bubeck C, Müllen K, Wegner G: Third harmonic generation in perylene derivatives ... 143

Schick C, Stoll B, Schawe J, Roger A, Gnoth M: Dielectric and thermal relaxations in low molecular mass liquid crystals ... 148

Roth HK, Baumann R, Bargon J, Schrödner M: Laser-induced conductivity in thin layers of poly(bisethylthio-acetylene) ... 157

Orczyk M, Sworakowski J: Phase transitions in single-crystalline polymerizing diacetylenes as seen by dielectric measurements ... 163

Author Index ... 167

Subject Index ... 168

Electromechanical behavior on thermoplastic with modified surface (Frost effect)

I. Chudáček and D. Slavinská

Department of Polymer Physics, Charles University, Praha, Czechoslovakia

Abstract: We have studied the surface and bulk deformation of polystyrene layers charged and heated to glass transition temperature T_g. Our layers have a modified surface obtained by ion bombardment. The experimental results lead to the conclusion that after charging and during heating, the surface charge produces an effect which, at room temperature, predominantly runs to surface energy with a consequent decreases of the surface tension. At temperatures near T_g, deformation of the polystyrene layer prevails. We have proposed a molecular model for the frost-effect.

Key words: Surface charge; roughness; surface tension; polystyrene

1. Introduction

During the development of thermoplastic recording in the period 1950—1970 and used in some electrophotographic processes a new phenomenon emerged involving the crazing of the thermoplastic layer [1]. It was found that when an ultrathin layer was added to the thermoplastic, an electrostatic charging and subsequent heating of this bilayer to the glass transition temperature T_g would result in a deformation of the surface. The suitable surface thin layer (of thickness up from only thousanths of atomic monolayer) of organic and inorganic materials can be produced on the thermoplastic by deposition from solution, by vacuum evaporation, or by ion bombardment. It is also possible to use cross-linking of the thermoplastic surface by short UV light, x-rays or electron beam to alter the polymer surface and produce the crazing phenomenon called the frost effect.

It is possible to record a charge pattern deposed on the thermoplastic surface [1]. At some critical temperature, which is near T_g, the frost pattern is developed only in the charged areas. This property is common with the thermoplastic recording. But for intentional frost-imaging, we must coat the thermoplastic layer with an ultrathin surface layer, the hardness of which is increased relative to the bulk of the thermoplastic. An example of a frost pattern is shown in Fig. 1. The electrostatic latent image is produced by x-ray irradiation of a modified surface of polystyrene which was shaded by metal objects.

It has already been pointed out from experimental results [1] that a large number of materials of very different physical and chemical properties are suitable for production of a useful surface layer. The surface charge is non uniform, in any case, because it is insufficient to charge every surface atom. For example, a polystyrene (PS) layer as thermoplastic, 25-μm thick, has a capacitance of approximately 80 pF and accepts a voltage up to 1000. The surface charge is therefore 10^{-7} C/cm^2 or 6.10^{11} electrons/cm^2. The electrons are spaced about 15 nm apart on the average. This charge non-uniformity is on the order of 10 nm, which is in [1] associated with surface islands, and it is amplified to a disturbance of 10 μm, which is the spacing between nearest cracks on the frosted surface. Attempts were made to observe disturbances on a clean surface by electronic microscopy at room temperature without charging and heating. Nicoll [1] came to the conclution that frosting does not occur at all without an ultrathin surface layer which can create a non-uniform surface tension. This tension produces the growth in size of the charge disturbances of 15 nm to the mechanical one of 10000 nm. The amplification of the charge disturbances by 2—3 orders of magnitude the absence of the surface layer does not occur.

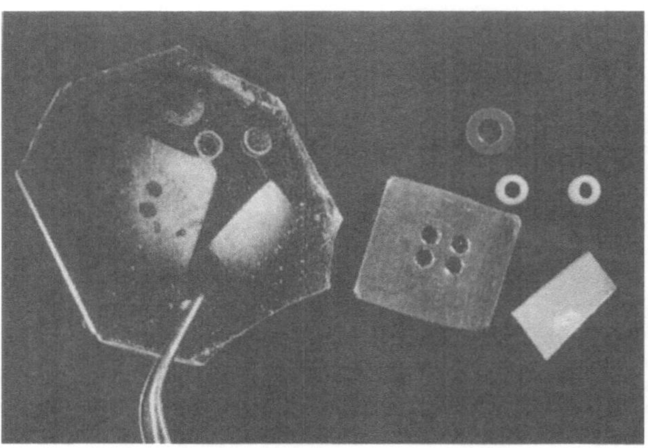

Fig. 1. A photograph of a frosted image on PS observed in an oblique light

Creesman [2] found experimentally that the frost effect would not occur below a critical value of the surface charge density or a threshold surface potential U_{sf}. He associates the creation of the frost at U_{sf} with the decrease of the surface tension γ of the thermoplastic to zero, in analogy with the electrocapillarity effect [3]. In this model, γ of a thermoplastic layer having a surface charge density q_s is given by

$$\gamma = \gamma_0 - \frac{1}{2} q_s U_s , \tag{1}$$

where γ_0 is the surface tension without charges on the surface. The second member on the right side of Eq. (1) is simply the electrostatic energy per unit area. At the threshold $\gamma = 0$ and because $U_s/d = q/\varepsilon$, where d and ε are the thickness and the dielectric constant of the thermoplastic layer, Eq. (1) becomes

$$U_{sf} = \sqrt{\frac{2\gamma_0 d}{\varepsilon}} . \tag{2}$$

Budd [4] presents a theoretical analysis of a surface deformation of an electrically charged thermoplastic film, where he uses spatial harmonics characterized by a wave number k to determine the growth rate w of the surface deformation. He assumes that the thermoplastic is an uncompressible viscose fluid without any surface layer. The predominant spatial frequency present in the frost pattern occurs at high growth rate w_{max}, which increases with the enhancement of the U_s. This approach was extended by Killat [5] and others [6] to take into account the tangentical stresses and to calculate the threshold surface potencial. In [7], the authors have measured the temperature dependence on U_s. A breakdown of this dependence occurs at some temperatures which correspond to the temperature dependence on the electrical conductivity of PS also occurs at this temperature (see Fig. 4). This temperature can be related to the glass transition of the given polymer determined by other properties (mechanical, dielectric, etc.).

The aim of the present work is to contribute to the explanation of the frost effect and relate it with the structure of the thermoplastic layer.

2. Experimental

In our experiment, we used aluminum sheets or glass slides as a substrate. The glass slides have transparent SnO_2 electrodes. We used as thermoplastic, low-molecular-weight PS (M_w = 8000), synthesized by the Institute of Macromolecular Chemistry of the Czechoslovak Academy of Sciences. The PS layers 10—20-µm thick were prepared by dipping into a viscous solution of PS in toluene (25 g of PS in 100 ml of toluene). The sample was pulled out from the solution, the solution allowed to drip off and the remaining solution from the sample border was drained off. Afterwards the samples were dried for a few hours.

The surface ultrathin layer was prepared by ion bombardment with atmospheric ions of a corona discharge in low vacuum. A homemade vacuum vessel was equipped with a grounded electrode, on which the metallic support of our sample lies. The second point-electrode is placed a few mm above the PS layer and is connected to the electrode with an AC supply of 6 kV, 10 kHz through an isolant (capacitance bound). The bombardment occurs at a pressure of 10 Pa and takes 1—10 s. The ion bombardment produces a very effective break of the chemical bonds in the macromolecules at the sample surface and, in this way, active macroradicals are produced. It is possible to observe a concentration of radicals by electron paramagnetic resonance after some seconds of bombardment. The lateral cross-linking of nearest polymeric chains may occur. At the same time the hydroxyl radicals which are created by the corona discharge because water is present in the vacuum vessel are adsorbed on macroradicals. The bombarded surface becomes hydrophylic under these influences.

We have determined the degree of the surface hydrophility by the measurement of the wetting angle. We have constructed a small vessel filled with water, where it is possible to inject the air bubbles which reach the immersed sample (placed horizontally with the analyzed surface down) from below. We use an optical microscope to measure the angle between the sample surface and the tangential plane of the bubble surface. In Table 1 there are data about the time duration of the bombardment t_b, the wetting angle α_w, and the stability of the frost, given by the time t_{fs} of its existence during a thermal contact with a metal desk which has a temperature of about T_g + 50°C. The thickness of the layer d = 15 μm.

One can see from Table 1 that at the beginning of the bombardment the break of the chemical bonds and the high speed attachement of OH radicals occur. This produces a large change in the wetting angle. After 5 s a sufficiently large amount of radicals is produced which permits the cross-linking between nearest chains and attachment of OH radicals to be diminished. The surface becomes not very hydrophilic. For longer times of the corona action on the PS surface some degradation of the present molecules occurs. The degradation products have larger hydrophility. At very long times of bombardment, the surface has a large degradation reflected by the yellow color of the sample, and the creation of the frost is no longer possible.

The measurement of the surface potential U_s was provided in a commercially available integrating electrometer Polystat, produced by JZD Jizera, Czechoslovakia. The measuring of U_s is given by integration of pulses induced by capacitance bounding with rotating charged sample [8] (see Fig. 2). We have arranged a new sample holder, which can heat the sample by an electric current up to 100°C. We measure this temperature by a thermistor, as shown in Fig. 3. The electrical current arrives at the sample holder (which rotates on a disc) by means of contact with brushes placed on the driving axis of the disc. We have calibrated the temperature gradient between the charged surface and the bottom of the glass support by a Cu-Constantan thermocouple under conditions which simulate the measuring ones. The measuring thermistor is integrated in a bridge circuit with a second referee

Table 1

t_b (s)	α_w (degree)	t_{fs} (s)
0	90—95	no frost
2	40—50	1
5	65—70	15
7	55—60	8
10	40—55	6
15	30—50	5
150	20—25	no frost
After 1 h of t_b = 5 s bombardment	75—80	10

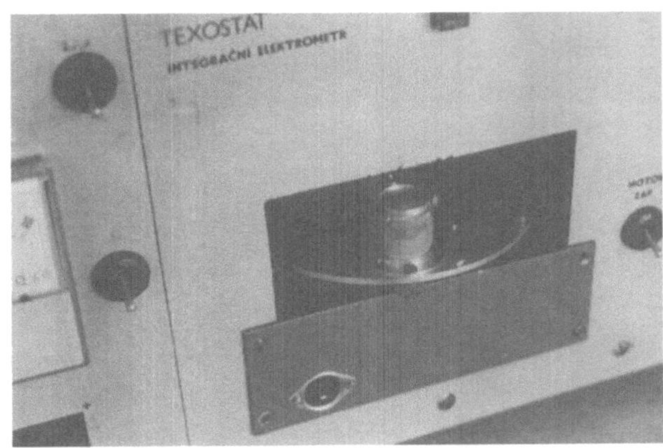

Fig. 2. An image of the sample holder, placed on the rotating disc of our integrating electrometer

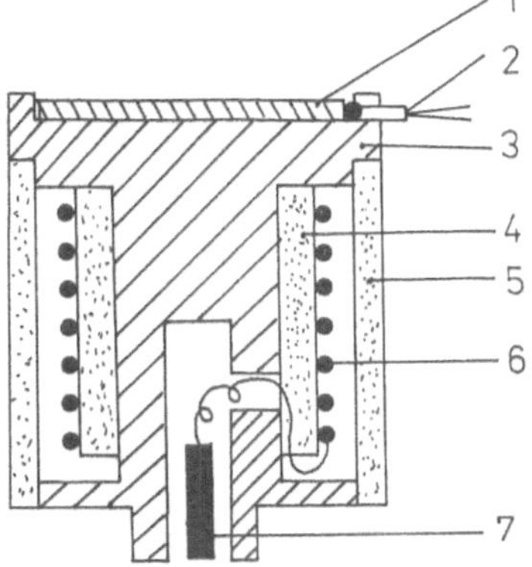

Fig. 3. A schematic picture of the heating sample holder. 1) sample; 2) thermistor; 3) heating body; 4), 5) teflon protective cylinder; 6) heating wire; 7) connector for heating current

thermistor, which is maintained constant at 0 °C temperature.

Samples placed on the holder were charged in the electrometer with a corona discharge from a needle to a potential in the limits +200 V to −1200 V. Shortly after charging, the change of their surface potentials was measured at different temperatures and different initial potentical U_{so}. In Fig. 4 is shown the temperature dependence of U_s for one U_{so}. In this figure is also shown the temperature dependence of the elastic modulus of our PS and the electrical conductivity [14]. The last, one was measured on the surface of our PS layer between two metallic electrodes, spaced 1.5 mm a part.

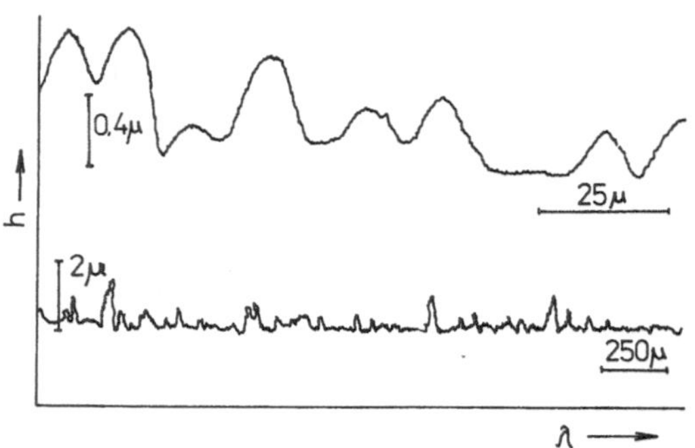

Fig. 5. The surface roughness of the frosted PS surface determined by the surfometer SF 200

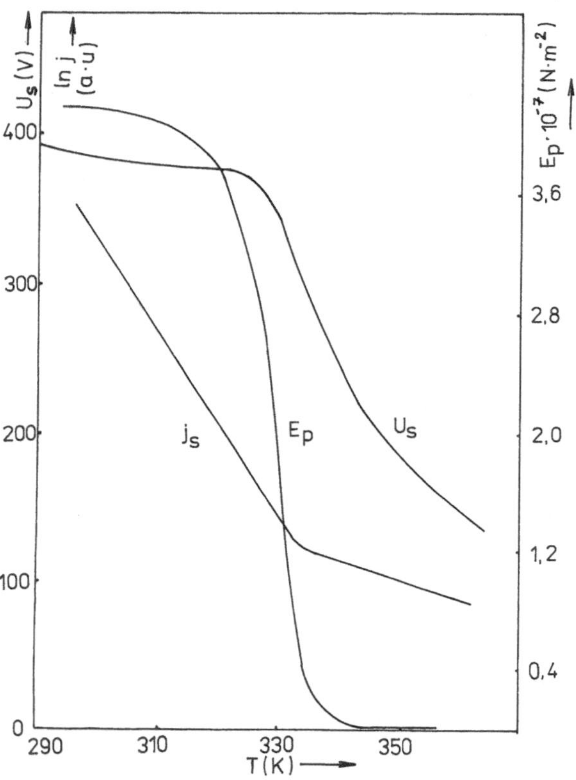

Fig. 4. The temperature dependencies of U_s, E_p, and surface current j_s *(on 1/T)*

a long path of the cursor with low horizontal and vertical sensitivity and a shorter path with higher sensitivities in Fig. 5. We can see the obvious quasi-periodicity of surface deformations on the frosted surface. In Fig. 7 there are shown the distribution of spacing between nearest deformations (reversal of the spatial frequency) and the distribution of the barrier heights between deformations. The average distance between deformations is 15 μm (67 l/mm). One can see in Fig. 6 that the similar modulation occurs on the non-frosted surface of our PS samples. This is an important experimental result from which follows that the frost effect is probably connected with some structure present in our

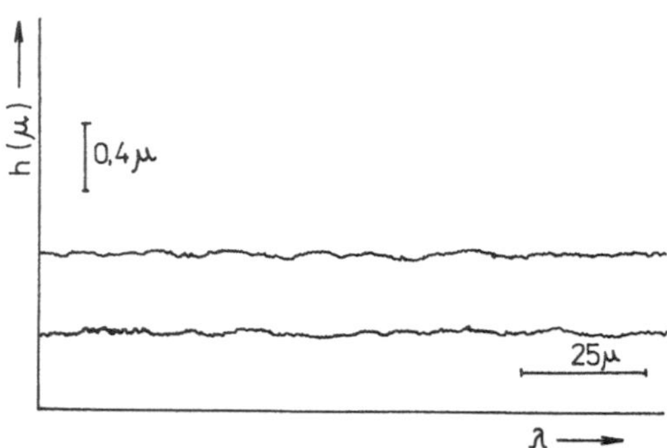

Fig. 6. The surface roughness of unfrosted PS surface

3. The pattern of the frosted surface

First, we examined the surface relief of the frosted PS surface by a commercially available surfometer SF 200 from Planer Products (GB). There are shown system before the charging and heating. The mean heights of the frost deformation are, before frost 30 nm, and after frost 600 nm. Therefore, the frost-effect produces a magnification of about 20 times as high as the initial surface modulation.

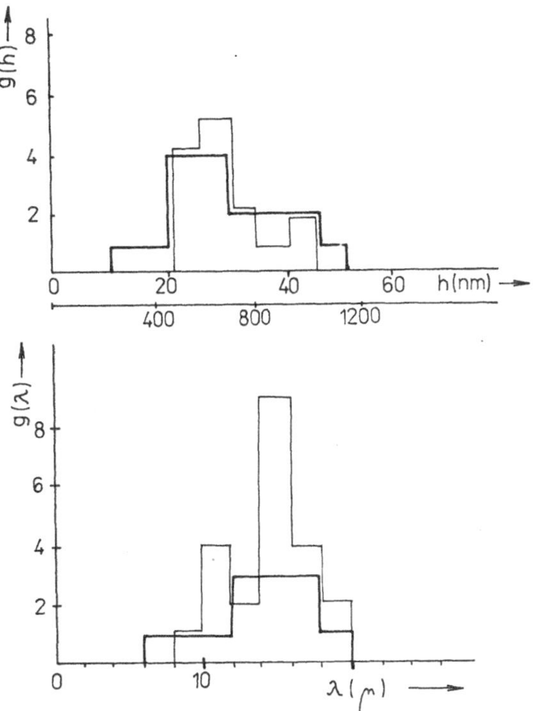

Fig. 7. The distribution of the barrier heights $g(h)$ and spacing $g(\lambda)$ between nearest surface deformations *before* frost and *after* frost

In Fig. 8 are the micrographs of frosted PS samples obtained by the metalographic microscope Neophot, Zeiss Jena. In Figs. 8a and 8b there are relatively thick thermoplastic layers (d = 25 μm). The heating and cooling rates are different for 8a

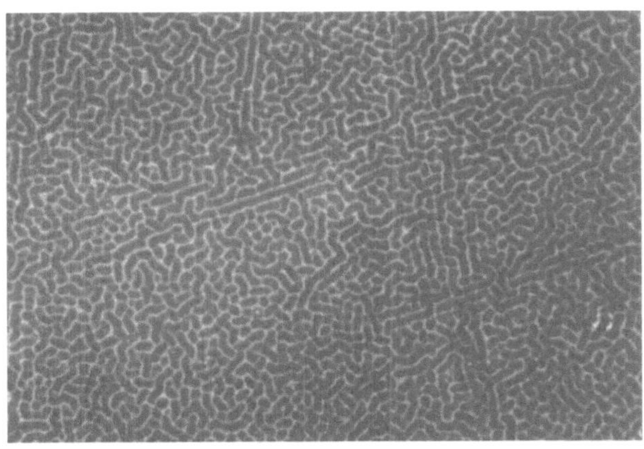

a

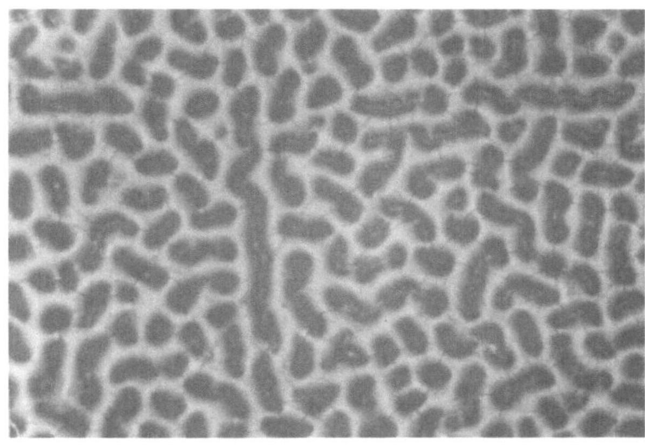

b

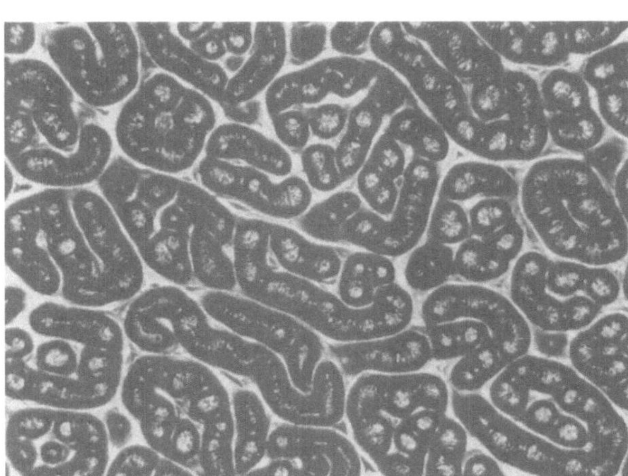

c

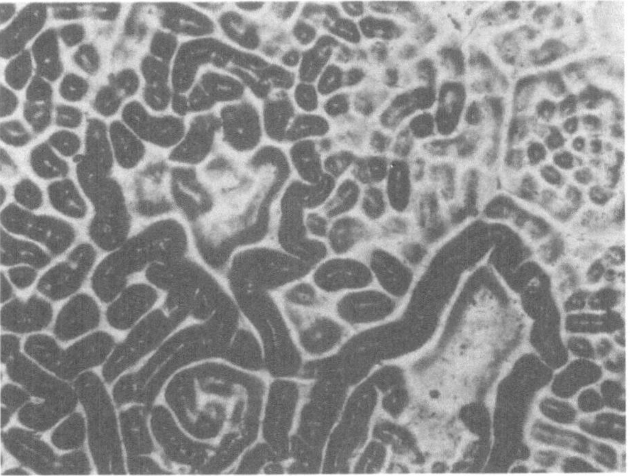

d

Fig. 8. Micrographs of frosted PS samples. a) thick PS layer, slow heating rate; b) thick PS layer, high heating rate; c) PS layer treated by thermal shocks; d) as with c) at the charge border

(5 K/s) and for 8b (15 K/s). In the case 8a, we have a very well developed frost structure with some bands of relatively different arrangement of the frost pattern. In the case 8b, we assume that the frosting is not completly spread out. Probably, the time interval in which the system is above the frosting temperature is very short. The mechanical relaxation of the surface does not occur completely. For this reason, we can observe local fracture lines of the original plane between frost deformations in Fig. 8b. In both cases the distribution of the nearest spacing between frost deformations L are practically the same. For the average, we have L = 7.7 μm (130 l/mm).

In Figs. 8c and 8d there are shown typical patterns of frosted surfaces before charging and heating, and thermally treated with thermal shocks. In Fig. 8d, we can observe the frosted surface of the border of the surface charge. The local symmetry of the frost deformations at this border is different from that in the middle of the frosted surface. In both cases the local symmetry in 8c and 8d differs from that in 8a and 8b. An experimental confirmation is that also the surface structure before charging and heating plays an important role in the frost pattern, but temperature behavior during the development of the frost also plays a significant role.

In Fig. 9 there are shown patterns of unfrosted (9a) and frosted (9b) PS surface (kindly provided by Dr. M. Stamm, Max Planck Institute für Polymerenforschung, Mainz, FRG). We have used their mathematical processing of an image coming from a phase interference microscope LOT/ZYGO, which can charaterize the sample surface with a lateral resolution power of about 1 μm and a height resolution of 0.6 nm [9]. We have recorded the frost at the surface charge border in 9b. One can clearly see the frost deformations which have expanded above the original surface during the development of the frost. Their geometrical distribution on the surface gives an averadge spacing between nearest deformation equal to L = 4.1 μm and an average height equal to 220 nm. If we compare this average spacing with values which we obtained with optical microphotography and with the surfometer, we can say that we have obtained lower values. This is probably due to the higher sensitivity of the image-processing in the last case. It is important that in 9a we have practically the same spatial distribution of the frost deformations for non frosted surface as in 9b for frosted surface (L = 3.7 μm). The determined mean height in the last case is 15 nm and from this

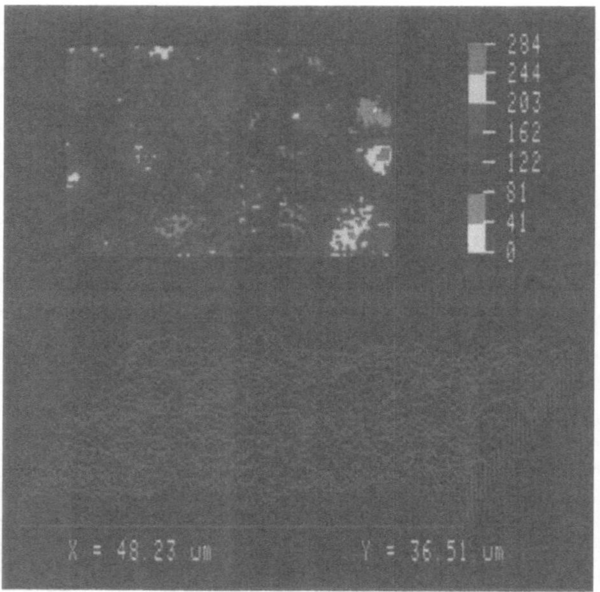

a

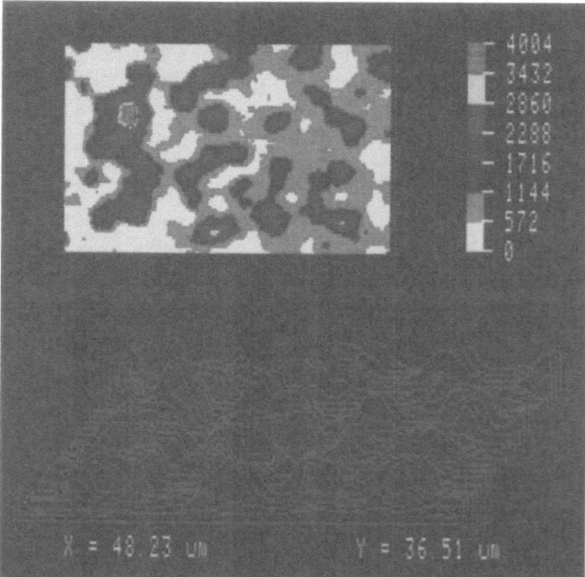

b

Fig. 9. Mathematically processed surface pattern generated from a phase interference microscope. a) unfrosted surface; b) frosted surface

value follows that frost produces a multiplication factor of 15 of the surface modulation; it is a little lower than that for the determination of the multiplication factor by the surfometer (20×).

The results presented lead to the conclusion that the frost effect is given, above all, by the existence of some structural fluctuations in the thermoplastic layer. These fluctuations can also produce fluctuations of the surface tension of our bilayer system. We can compare the frost pattern with the pattern obtained by the imaging of surface phase separation occurring during a spinodal decomposition. The spinodal decomposition occurs during various thermal treatments, cooling or heating, when a stable single phase may be made unstable with respect to a mixture of two phases [10]. In the spinodal decomposition a high degree of connectivity among particles of each phase is typical, which is also observed in frost pattern (see Fig. 8). In polymer systems the spinodal decomposition is related to the polymer dispersity [11, 12]. The frost-effect plays the role of noise in thermoplastic recording of holographic pattern on a suitable thermoplastic such as PS. The authors of [11] have taken a monodisperse PS and provided cycles of thermoplastic recording, of course, including erasing of the surface relief between thermoplastic cycles. In the beginning there is no observed frost. After 5—7 cycles of recording erasure the frost appears again. The authors assume that this is caused by the influence of the corona discharge, where the ion bombardment produces the scission of the polymer chains and, thus, the change of the system dispersity.

5. The temperature dependence on the threshold potential for frost

The experimentally determined temperature dependences $U_s(T)$ for different U_{so} are shown in Fig. 10. In the beginning, U_s monotically decreases up to a temperature where a break in this dependence occurs. At this temperature the frost appears (visible to the naked eyes) and it can be preserved by a cooling down of the sample (see also [7]): The corresponding values of the frost appearance are U_{sf} and T_f. The relation between U_{sf} and T_f for different U_{so} gives us the curve of the temperature dependence of the threshold surface potential. We have optically determined the temperature interval t_{fs} of the frost stability above T_f during heating of the sample with a rate of 10 K/s. If we add t_{fs} to T_f for a given curve $U_s(T)$, we determine the high temperature border of the existence of frost. Above this border the erasing of the frost occurs, probably produced by segmental motions of polymeric chains.

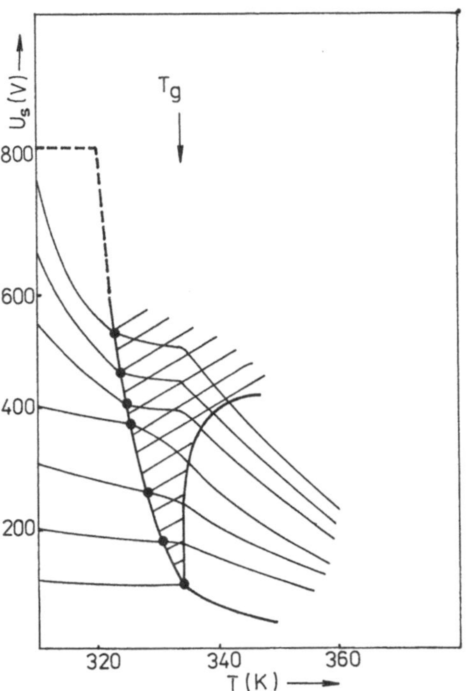

Fig. 10. The temperature dependence of U_{sf} (black points). The lined area is the area of the frost stability

The frost-effect is a surface phenomenon produced by the charge placed on the sample surface. We consider that the surface charges have no free movements on the sample surface. Their movements are connected with the mechanical deformations of the surface or of the bulk of the sample. We assume that the injection of the surface charge from surface into the bulk, which increases with temperature, does not affect the frost-effect.

Elementary charge generates an effect during the charging — heating process. At some temperature T an enhancement ΔT of the temperature produces the effect of the electron ΔL_e. This effect will be transformed in the enhancement of the surface energy ΔL_s and the mechanical work ΔL_d related to the sample deformation. We have for one electron:

$$\Delta L_e = \Delta L_s + \Delta L_d \,, \tag{3}$$

where

$$\Delta L_e = \frac{eU_s(T)\Delta d}{d} \tag{4}$$

$$\Delta L_s = \gamma \Delta S_e \tag{5}$$

$$\Delta L_d = E_p \frac{\Delta d}{d} S_e \Delta d \, , \tag{6}$$

where Δd is the deformation of the sample thickness d, E_p is the elastic modulus of our PS, ΔS_e is the increase of the surface corresponding to one electron S_e.

For S_e, we have:

$$S_e = \frac{e}{q_s} = \frac{ed}{\varepsilon U_s} \, . \tag{7}$$

If we put (4)—(7) in Eq. (3), we obtain

$$U_s^2 = \frac{G \cdot \gamma}{\varepsilon} + \frac{E_p}{\varepsilon} \cdot \Delta d \cdot d \, , \tag{8}$$

where $G = \dfrac{\Delta S_e / S_e}{\Delta d / d} \gg 1$ as follows from experimental results.

For the beginning of the heating process, at low temperature the thickness deformation is very small $\Delta d \to 0$. In Eq. (8) the first coefficient on the right side becomes very important. If we take into account the electric field dependence of γ, as in Eq. (1), we obtain

$$U_s^2 \sim \frac{G\gamma}{\varepsilon} = \left(\gamma_0 - 1/2 \frac{\varepsilon U_s^2}{d} \right) \frac{G}{\varepsilon} \, .$$

It follows that

$$U_s^2 \sim \frac{2\gamma_0 d}{\varepsilon} \, ,$$

which is another form of Eq. (2). In the last equation, we can see that the temperature dependence of $\gamma_0(T)$ is determined by the temperature dependence of U_s. This $U_s(T)$ function, as we know, is given by the temperature dependence of the bulk conductivity of the thermoplastic and the temperature dependence of the injection of charges from the surface into the bulk. With the increase of T occurs the decrease of U_s, and in accordance with the last equation the decrease of γ_0. Equation (8) decreases the value of the first coefficient; the second coefficient also depends on temperature through the temperature dependence of $E_p(T)$ (see

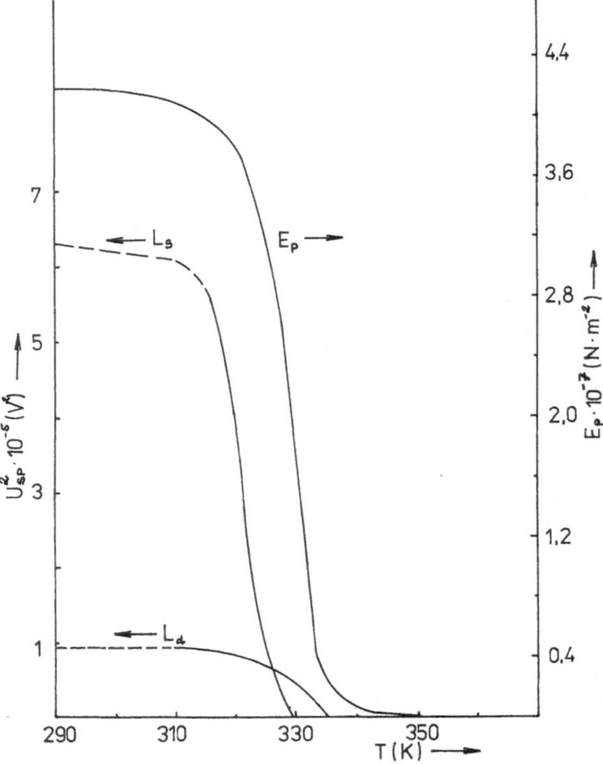

Fig. 11. The temperature dependence of E_p, L_s, and L_d

Figs. 4 and 11). In Fig. 12, we have shown the experimental dependence of U_{sf} on E_p, starting from the temperature dependencies of both values. The dependence of U_{sf}^2 on E_p is linear only for temperatures higher than 328 K, e.g., 6 K below T_g = 334 K. We can decompose $U_{sf}^2(T)$ on the part given by the first component (surface energy L_s) and by the second component (deformation energy L_d). For comparison, we have shown $L_s(T)$ and $L_d(T)$ in Fig. 12 together with $E_p(T)$. If we do an extrapolation to room temperature, we can determine that the effect of the electric charge is transformed approximately to 90% as surface energy, and to 10% as deformation energy. When the temperature increases the ratio L_s/L_d diminishes and becomes unity at 326 K, e.g., 8 degrees below T_g. This can be interpreted as the surface tension having a stronger decrease with temperature than does the elastic modulus. From our interpretation, it follows that the condition $\gamma = 0$ is insufficient in a deformation of the thermoplastic layer.

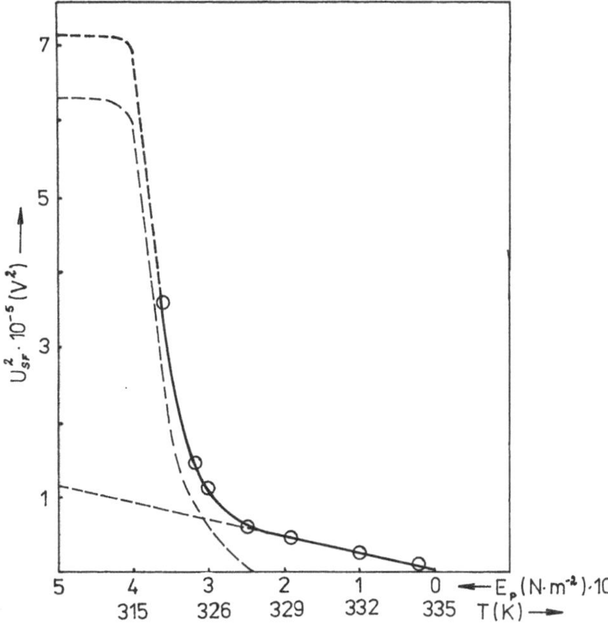

Fig. 12. The dependence of U_{sf}^2 on $E_p(T)$ and decomposition in the surface and bulk part (see text)

At temperature near the glass transition becomes predominant the mechanical deformation of the bilayer system. We propose now, following the presented experimental results, a model for the frost-effect. We have assumed that the surface charge is bounded (probably by non-chemical bonds) by the surface structure of the bilayer system. Between them two different Coulombic interactions occur:

A) Repulsion between charges of the same polarity bounded to the free surface. These forces are not completely compensated, because the charges do not lie in one plane. For charges placed above the mean statistical plane (see Fig. 13) the rest of surface charges repulses more above these charges (see Fig. 14b). For charges placed under the mean statistical plane, the repulsion forces also repulse more under these charges. The results of this interaction are the diminishing of the surface density, the increasing of the thickness of the surface layer, and the subsequent decrease of the surface tension.

B) Attraction between charges of opposite polarity placed at the two electrodes of the sandwich-like sample: the support electrode made of metal or transparent conductive material deposed on glass, and the virtual electrode made by surface charge on our bilayer system. The result of this interaction is the compression of our bilayer.

In Fig. 14, we have schematically shown our model. In 14a, we see the profile of the surface layer, obviously 50-Å-thick [14], having a free surface fluctuation of about 30 Å [14] and an interface fluctuation between surface layer and thermoplastic about 20 Å [14]. The quasi-periodical modulation of the unfrosted surface layer is probably given by the relaxation of the surface tension which take place during the preparation of the sample. In 14b, we see the interaction analyzed in A) and B); it is more effective with increasing of the surface potential or the temperature. These interactions produce mechanical stresses in the surface layer and under it. At some critical point, in connection with the present surface defects, the rupture of the surface layer occurs. Then the relaxation of local stresses by

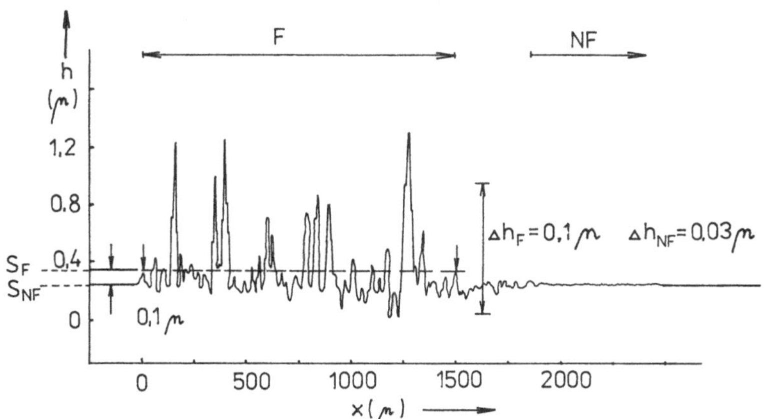

Fig. 13. The roughness at the border of unfrosted and frosted PS surface determined by surfometer SF 200 for one sample. S_{NF}, S_F are the statistical surface planes for unfrosted and frosted area, h_{NF}, h_F are the main barier heights

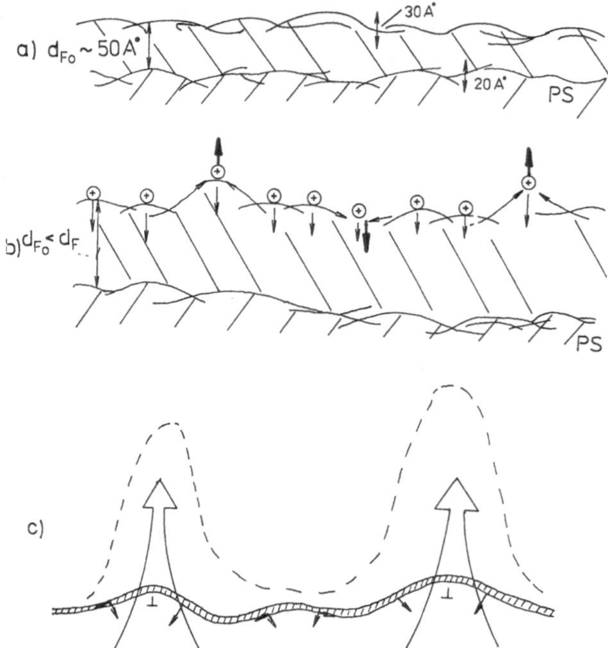

Fig. 14. The schema of the frost effect. a) surface thin layer before charging, b) surface after charging, c) compression and relaxation by plastic deformation

local plastic deformations takes place, which results in the final shape of the frosted surface (14c).

Conclusion

We have seen in the introduction that there are two different concepts about the frost-effect that occurs at the surface of a thermoplastic:

i) The first, initially proposed by Nicoll [1], claims that the frost-effect occurs only when the surface of the thermoplastic is modified in such a manner that it becomes harder than the thermoplastic bulk. The main role of this ultrathin surface layer is to amplify the spatial heterogenities of surface charge to observed frost pattern.

ii) The second, initially proposed by Creesman [2], explains the role of the surface charge on the thermoplastic layer by diminishing the surface tension. The thermoplastic is treated as a viscous incompressible liquid.

Our experimental results favored concept i). We have treated the thermoplastic as a solid glass which can be deformed. We have also observed a high decrease in the surface tension during the development of the frost on the charged surface. At temperature near T_g the mechanical deformation of the sample becomes predominant, just before frosting of the surface.

We have demonstrated that the surface patterning by the frost-effect must be in relation to the internal structure of the thermoplastic. The topology of the frost is very probably connected with the spinodal decomposition of our metastable thermoplastic occuring during the thermal history of the polymer used.

Acknowledgement

The authors are indebted to Dr. M. Marvan, Dr. K. Ulbert, and Dr. M. Stamm for stimulating discussions, and to J. Hejhálek for technical assistance.

References

1. Nicoll FH (1964) RCA Rev 25:209
2. Creesman PJ (1963) J Appl Phys 34:2327
3. Bikerman JS (1958) Surface Chemistry Academic Press New York, p 450
4. Budd HF (1965) J Appl Phys 36:1613
5. Killat U (1975) J Appl Phys 46:5169
6. Matthies DL, Johnson WC, Lampert MA (1975) J Appl Phys 46:2956
7. Ulbert K, Chudáček I, Slavinská D (1982) Polymer Bull 8:179
8. Ulbert K CS Patent AO 215536
9. Biegen JF, Smythe RA (1968) SPIE O/E LASE '88 Symposium Jan 1968, Los Angeles
10. Hüttenbach J, Stamm M, Reiter G, Foster M (1990) Submitted to Langmuir
11. Cahn JW (1965) J Chem Phys 42:93
12. Lee TC (1974) Appl Optics 13:888
13. Gravel LL (1975) Appl Optics 14:2054
14. Sorm F (1982) Institute of Molecular Chemistry, Prague, private communication
15. Stamm M (1990), in this proceedings

Authors' address:

Dr. I. Chudáček
Department of Polymer Physics
Charles University
V. Holešovičkách 8
18000 Praha, Czechoslovakia

Discussion

WÜNSCHE:
How does the dimension of the layer influence the frost effect?

CHUDACEK:
The thickness of the thin surface layer controls the appearance of the frost effect, which occurs only in a certain interval of thicknesses. The particular values of this interval depend on the nature of the surface layer. Examples: polymeric overcoating 10—200 Å, metal coating 0.01—20 Å.

STAMM:
How do the properties of the polymer surface and, in particular, surface roughness influence the frost-effect?

CHUDACEK:
In our experiments, we have observed frost-effect only on polystyrene coated by polyvinylalcohol, or on modified polystyrene surface by ion bombardment. In both cases, the sample surface is harder than the thermoplastic bulk. In short, we can say that the frost-effect magnifies the surface roughness.

STAMM:
With incompatible polymer films like polystyrene on poly-parabromostyrene, we also observe, above the glass transition temperature, the formation of lateral structures. We believe this is an effect due to minimization of surface tension. Do you believe that the frost-effect is also influenced by the compatibility of materials, and is it only observed after changing?

CHUDACEK:
In our experimental situation the frost-effect occurs on charged surface of bilayer system during the decreasing of the surface tension. But the effect is generated, above all, by local stresses present and their relaxation by local plastic deformations. It is possible to imagine these conditions without the presence of charged surface, for example, in an interface of incompatible polymers.

GERHARD-MULTHAUPT:
Could the "amplification" of the surface deformations during frost formations be explained by the feedback mechanism, which is caused by the inverse square dependence of the deforming stress on the local layer thickness?

CHUDACEK:
In fact, from Eq. (8) for temperature near T_g, it follows that the mechanical stress perpendicular to the surface is

$$\sim \sigma \Delta d \frac{U_s^2}{d^2},$$

and a feedback mechanism takes place (local plastic deformations). The problem is how to interpret the thickness d.

Deformation of semicrystalline and molten polyethylene the role of entanglements

E. Paul, B. Heise, W. Schrodi, and H. G. Kilian

Abteilung Experimentelle Physik, Universität Ulm, Ulm, FRG

Abstract: Pressure-strain curves measured for high- and low-density polyethylene under rapid and different strain rates at different temperatures are described with the aid of the adequate van der Waals network equation of state. The homogeneous melt behaves like a quasi-permanent network with the same entanglement concentration as deduced from the plateau-modulus. Entanglements can be identified by large fluctuations approximating the ones of chain ends. It is demonstrated that entanglements cannot be annihilated during crystallization. Entanglements are squeezed out of crystals so that their configuration is related to the primary colloid-structure. Pressure-strain curves of semi-crystallized samples are described by using the same parameters as in simple extension. Deformation mechanism in semicrystalline polymers are deformation-mode invariant. Entanglements should be "condensated" during the stretch so that the effective chain-length in the crystal-cluster network exceeds at least the one observed in the melt.

Key words: Polyethylene; large deformation; entanglement-network; van der Waals-network model

Introduction

Deformed under high strain rates, polymer melts were found to behave like networks. Entanglements were assumed to operate as quasi-permanent junctions. This is the reason why entanglements are believed to influence the modulus of networks [1—23].

Due to these findings, we postulate that entanglements cannot be annihilated during crystallization. Because of being lattice-incompatible, entanglements must be squeezed out of crystals into defect layers or amorphous regions [3, 2, 3]. The entanglement configuration must be determined by the colloid structure.

Moreover, because entanglements cannot be built into a crystal lattice, their configuration must be changed during deformation when the superstructure is altered [3, 16—36].

To prove these ideas samples were rapidly pressed to sheets of a thickness between 10—100 μm at different temperatures for different strain rates. With this method it is possible to study the pressure-strain behavior of melts and of semicrystalline systems.

To find out whether entanglements are annihilated during crystallization and forced into a special configuration, semicrystalline samples were pressed at a temperature so as to guarantee deformation-induced melting of the entire crystal ensemble. In the first moment, the configuration of entanglements is the same as in the semicrystalline system. This "non-equilibrium network" needs time to relax into equilibrium. For rapid deformations, a network with a non-equilibrium configuration of entanglements should determine the mechanical response.

The van der Waals network concept delivers a description of all these experiments. It will be shown that this description allows to discriminate between chemical junctions and entanglements and to get information about the configuration of a "non-equilibrium entanglement network".

Theory

First, we recall relationships with the aid of which large deformation in networks and in semicrystalline systems can be described [3, 21].

The structure of van der Waals networks

It was shown elsewhere that rubbers may be considered to behave like a conformational van der Waals gas network with weak interactions [37, 38]. It is for this reason that different modes of deformation can fully be interpreted [39]). In every mode of deformation the network is characterized by its mean chain length as number of stretching-invariant units and by the parameter of global interactions.

The maximum strain λ_m is the first van der Waals network parameter defined by the mean chain-length according to

$$\lambda_m = \sqrt{y_s}; \quad y_s = \frac{M_c}{M_u} . \tag{1}$$

M_c is the mean molecular weight of the chains, M_u is the molecular weight of the stretching-invariant unit ("entropy-invariant unit") which seems practically not to depend on temperature. The number of stretching-invariant units per chain, y_s, may therefore be taken as a strain invariant structure parameter.

The second van der Waals parameter a characterizes global interactions. In a phantom-chain network, interaction has to occur across junctions. Because of the stochastic nature of exchange of energy and momentum, junctions have to fluctuate [33]. The van der Waals interaction parameter is related to these fluctuations. This explains the observation that the interaction parameter decreases with increasing functionality of chemical crosslinks [40].

The equation of state

Calling the strain in perpendicular direction to the pressure applied λ, the van der Waals equation of state for equibiaxial deformation of networks defines the nominal force f as [41]

$$f = GD\left(\frac{1}{1-\eta} - a\sqrt{\phi}\right) , \tag{2}$$

whereby

$$D = \lambda^4 - \lambda^{-2}; \quad \phi = \frac{1}{2}\left(2\lambda^2 + \frac{1}{\lambda^4} - 3\right), \tag{3}$$

with [42, 43]

$$\eta = \sqrt{\frac{\phi}{\phi_m}}; \quad \phi_m = \frac{1}{2}\left(\lambda_m^2 + \frac{2}{\lambda_m} - 3\right) . \tag{4}$$

The network parameters λ_m, a and M_u are mode invariants. For this reason, the potential Φ_m can be formulated in the mode of simple extension (Eq. (4)).

With the aid of Eq. (1) the modulus may be written as

$$G = \frac{\rho RT}{M_c} = \frac{\rho RT}{M_u \lambda_m^2} . \tag{5}$$

Thus, it is derived that the size of λ_m influences both the relative course of the stress-strain curve and the absolute stress. The interaction parameter influences the relative course of the stress-strain curve only. This is the reason why both parameters λ_m and a can mostly be determined with a relative high degree of accuracy. This procedure is simplified, because M_u is determined by the chemical structure of the stretching-invariant unit so as to be identical for differently crosslinked networks of the same type.

Semicrystalline systems

One has to be aware that, during deformation, crystallites operate as active fillers. They undergo plastic deformation or melting and recrystallization. In polyethylene, plastic deformation is made by shear-sliding within (110)-planes and equivalents (c-axis-slip) [21]. Due to the few planes available, extended cluster aggregates must operate as smallest units which are transformed according to the affine law. These "equivalent subsystems of deformation" are supposed to be isotropically linked in their environment.

Processes running off within the subsystems are controlled by the primary colloid structure (crystallinity, mean size of crystals, entanglements, and tie-molecules). Yet, because of being confronted with complicated and cooperative processes, one has no chance for developing a detailed model. For understanding large deformations of semicrystalline systems, we must, therefore, rest content with a semi-phenomenological conception. It is important that parameters of the primary colloid structure enter into this approach.

The primary structure of LDPE [44, 45, 46, 67, 46]

Short-chain branchings of LDPE are lattice incompatible units; they are squeezed out of crystals. It is then the broad CH_2-sequence-length distribution which determines the thickness distribution of the *c*-sequence mixed crystals (inhomogeneous microphases: IMPs) as a typical feature of eutectoid multi-component systems. For a random distribution of *nc*-units, the means thickness of the IMPs is given by

$$\langle y_p \rangle = \frac{x_c}{x_{nc}} + y(T) , \tag{6}$$

where x_{nc} is the mol fraction of the *nc*-units

$$x_{nc} = \frac{n_{nc}}{n_c + n_{nc}} ; \quad x_c + x_{nc} = 1 ; \tag{7}$$

n_c and n_{nc} are the mol numbers of the co-units.

It is was found that the effective mass fraction of solid filler w_p has to include the defect layers of IMPs [26, 3]. This mass fraction is defined by

$$w_p = x_c^{y(T)-1}[y(T)x_{nc} + x_c] , \tag{8}$$

while the "two-phase" degree of crystallinity is written as

$$w_c = \left(1 - \frac{A}{3}\right) x_c y(T)^{-1}[(y(T) - y_k)x_{nc} + x_c] < w_p , \tag{9}$$

with

$$y_k = \frac{B}{3 - A} ; \quad B = B_1 \frac{x_c}{x_{nc}} + B_2 . \tag{10}$$

A, B_1, and B_2 are constants; $y(T)$ is the thickness of the smallest IMPs which are just stable at the coexistence temperature T:

$$T = \frac{T_m \left(1 - \dfrac{2\sigma_e}{N1}\right)}{N_2} , \tag{11}$$

where

$$N_1 = \left(1 - \frac{A}{3}\right) \Delta h(T)(y(T) - y_k) \tag{12}$$

$$N_2 = 1 + \frac{RT_m}{N_1} \left[\ln \left(\frac{y(T)}{\Delta y/2 + 1} \right) - \ln \left(\frac{x_y^{(m)}}{x_y^{(c)}} \right) + \frac{(y(T) - 1)}{2} \chi^{(m)} \right] ; \tag{13}$$

σ_e is the molar free surface enthalpy per CH_2-unit and $\Delta h(T)$ the molar melting enthalpy per CH_2-unit. $x_i^{(i)}$ is the mol fraction of the CH_2-sequences of the length y in the melt (m) or in the mixed crystal (c). $\chi^{(m)}$ is the interaction parameter in the melt. T_m gives the asymptotic melting temperature of an infinitely large extended *c*-sequence crystal. $x_y^{(m)}$ and $x_y^{(c)}$ are the mol fractions of CH_2-sequences of the length y in the amorphous or in the crystalline regions, respectively. w_p and $\langle y_p \rangle$ can be calculated with the aid of the above equations by adjusting $y(T)$ whereby the calculation has to reproduce $w_c(T)$ data as obtained by WAXS- or DSC-measurements. For more details the reader is referred to the original publications [31, 32, 33, 38—40, 46].

It is now a significant matter that eutectoid segregration of CH_2-units determines the primary colloid-structure. There are many reasons for postulating that this also happens to occur in largely strained samples.

Deformation

The above suggestion is justified by the description of strain-induced crystallization [48, 3]. Even at largest strains, *nc*-units are not built in the crystal core. The thickness distribution of IMPs is the same as in the unstrained system. According to thermodynamics, melting occurs selectively and consecutively across a broad temperature range.

The degree of crystallinity and the mean thickness of the amorphous layers (the last one being uniquely related to the mean thickness of the CH_2-sequence mixed crystals) depend uniquely on the actual temperature. For non-isothermal deformation of eutectoid copolymers, melting of defined fractions of IMPs should have to occur. The pressure-strain behavior is therefore strongly affected by partial melting. For describing non-isothermal deformation $y(T)$ must be correctly adjusted. The temperature at which deformation is running off is then deduced with the aid of Eq. (11) and the parameters listed in Table 1.

Table 1. Thermodynamic and structural parameters. L 1840 D [41, 46]

$A = 0.15$; $B_1 = 0.8$; $B_2 = 50$; $\chi^{(m)} = -0.01$
$T_m = 415$ K; $2\sigma_e = 8800$ J mol^{-1} CH_2^{-1};
$\Delta h(T_m) = 3971$ J mol^{-1}
$\Delta C = 5.88$ J mol^{-1} K^{-1}; $x_{nc} = 0.033$
$M_u = 68$ g mol^{-1}; $\rho(T)$ according to [24]

The network transformation

Two phenomena seem to control the deformation process in semicrystalline polymers [21], one of them is the operation of a crystal-cluster network with an effective chain-length which increases with increasing strain:

$$\langle y_m \rangle = \lambda_m^2 = a \langle y_p \rangle \left(\frac{1-u}{u} \right) ; \quad u = \left(\frac{w_p}{\lambda} \right)^{1/3} .$$

Having the equivalent subsystems of deformation isotropically linked in the environment is the reason for using $w_p(T)^{1/3}$, in Eq. (14) instead of $w_p(T)$. The parameter a scales how the effective chain length is related to the mean thickness of the amorphous layers. a can only be considered as an invariant structure parameter if the effective mean chain length is always related in the same manner to parameters of the primary colloid structure (w_p, $\langle y_p \rangle$). When the parameter a is constant, not depending on temperature for example, this must be taken as a manifestation of self-similarity of deformation mechanism in crystal-cluster networks.

The intrinsic strains

The second phenomenon is that a semicrystalline system is heterogeneously strained. Due to being composed of soft non-crystallized regions and solid crystals in series connection, the soft amorphous regions become heavily overdrawn. By plastic deformation of crystals this overdrawing is continuously reduced with increasing strain. This can be described by formulating the intrinsic strain in the rubbery amorphous layers as

$$\lambda_i = \frac{\lambda - u}{1 - u} ; \quad u = \left(\frac{w_p}{\lambda} \right)^{1/3} . \tag{15}$$

For $\lambda = 1$ this relation equals the Bueche-equation [21, 49, 50, 51]. The degree of overdrawing is supposed to disappear at infinitely large strains so that the crystal ensemble here displays "ideal plastic deformation":

$$\lim_{\lambda \to \infty} \lambda_i = \lambda . \tag{16}$$

Plastic deformation of crystals enforced from outside constraints come into play. Their strain-dependent configuration mirrors plastic deformation of the crystal ensemble which is described be defining the plastic strain parameter, λ_c. From the evident relationship

$$\lambda = w_p^{1/3} \lambda_c + (1 - w_p^{1/3}) \lambda_i , \tag{17}$$

λ_c is deduced to be given by

$$\lambda_c = \frac{\lambda - (1 - w_p^{1/3}) \lambda_i}{w_p^{1/3}} . \tag{18}$$

The equation of state

It is now assumed that mechanic equilibrium is bound to the condition

$$\langle f_{\text{crystals}} \rangle_{\lambda_c} = \langle f_{\text{amorph}} \rangle = f , \tag{19}$$

where $\langle f_{\text{crystals}} \rangle_{\lambda_c}$ is the Hooken-stress exerted on each crystal after the plastic deformation λ_c. According to Eq. (18), the pressure-strain pattern of the semicrystalline system is uniquely related to the deformation of the amorphous regions. The nominal force within these regions should be described with the aid of the van der Waals equation of state, in the mode of uniaxial compression written as

$$f = G_n D(\lambda_i) \left[\frac{1}{1 - \eta(\lambda_i, \lambda_m)} \right] , \tag{20}$$

with

$$\eta(\lambda_i, \lambda_m) = \sqrt{\frac{\phi(\lambda_i)}{\phi(\lambda_m)}} ; \tag{21}$$

D and ϕ are the same as defined in Eqs. (3) and (4) with λ replaced by λ_i. The modulus, written as

$$G_n = \frac{\rho RT}{M_u \lambda_m^2} ; \quad \lambda_m = \sqrt{\langle y_m \rangle} , \tag{22}$$

is a consequence of Eq. (14), i.e., that the density of effective crosslinks in crystal-cluster networks decreases with increasing strain. Because crystals operate as multi-functional junctions, the van der Waals interaction parameter a disappears [34].

Experimental

Materials

LDPE was investigated, the thermodynamic and structural parameters of which are listed in Table 1 [52, 53]. The data of fractions of LDPE of varying molecular weight are collected in Table 2.

Uniaxial compression

The material was covered with aluminium foil sprayed with polytetrafluor-ethylene powder so as to control whether shrinkage occurs after deformation and solification. In a hydraulic press these sandwiches were pressed mostly within 20 s at different temperatures, applying pressure in the range up to 120 kN. A sketch of the device is shown in Fig. 1.

To deduce the nominal force, the dimensions of the unstrained samples must be known. One has to control how these dimensions change by thermal expansion.

Table 2. Fractions of polyethylene

Sample	HDPE1	HDPE2
$\langle M_n \rangle$ 10^3 mol^{-1}	66	23

Acknowledgement: We are very much obliged to Dr. Boehm and Dr. Enderle of the Hoechst GmbH for supplying us with the samples investigated in Table 2

Results and discussion

Deformation of the homogeneous melt

It is to be seen from Fig. 2 that the pressure-strain curve of LDPE in the melt can fairly well be described with the aid of Eq. (2). The parameters used are collected in Table 3. Because of having an invariant set of network parameters, the existence of a quasi-permanent entanglement network is beyond any doubt. Within the time-window of these experiments, and within the limits of accuracy, the strain rate does not modify the stress-strain pattern. Deformation-induced heating of the samples should be small because of the large distance to the glass transition range [54].

It is a significant finding that disentangling cannot be enforced, even at the largest stresses applied ($f \approx$ 0.4—0.5 GPa). This agrees with rheological measurements [1, 2]. The modulus calculated with the relationship defined in Eq. (5) leads to a plateau modulus which is in accord with experiments [49].

How the interaction parameter depends on the crosslinks functionality φ was predicted with the aid of the empirical relationship [34]

$$a = \frac{K}{\varphi^\beta} ; \quad K = 0.6 ; \quad \beta = 0.7 . \tag{23}$$

Representative theoretical values are listed in Table 4. Values in the range a = 0.45—0.59, as deduced from our experiments, lead to the very interesting suggestion that entanglements fluctuation approximates the one of chain ends (φ = 1 in Table 3).

A model which explains this effect is two entangled loops, as shown in Fig. 3. Momentum and energy being exchanged nearly in the same manner between entangled and nonentangled chain

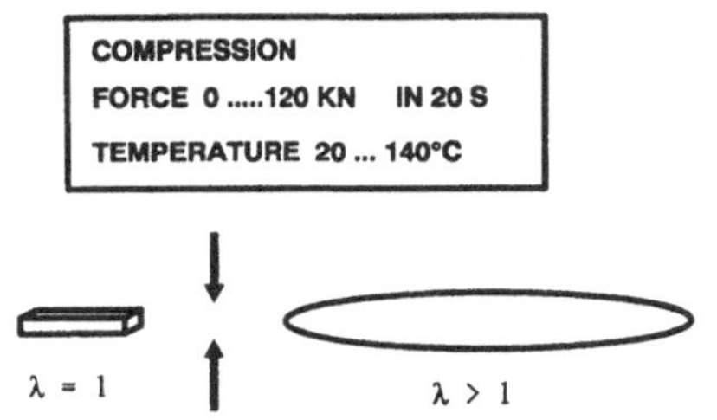

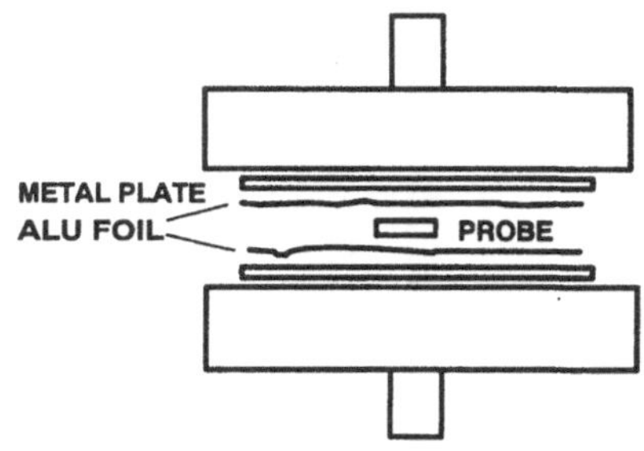

Fig. 1. Sketch of the experimental set-up

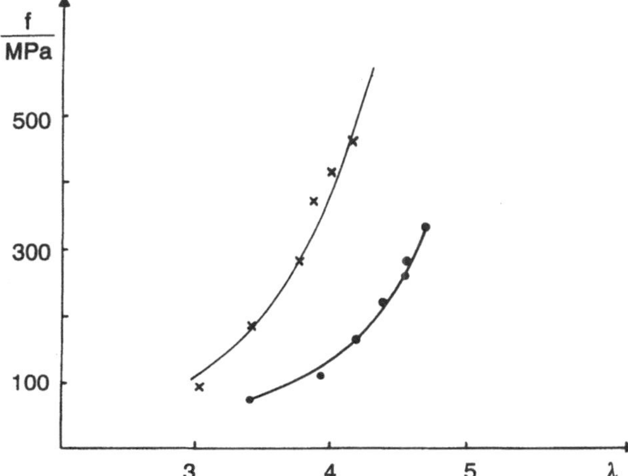

Fig. 2. Pressure against strain for LDPE at T = 388 K (●) and T = 378 K (*). The deformation was carried out within 20 s. The solid lines are theoretical (Eq. (2)). Parameters used are listed in Table 3

Table 4. Interaction parameter a of van der Waals networks

$K = 0.6$; $\beta = 0.7$

φ	1	3	4	6	10	100	1000
a	0.6	0.28	0.22	0.17	0.12	0.024	0.005

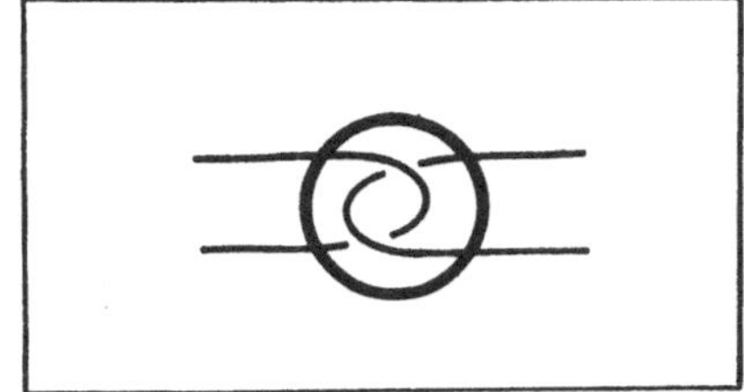

Fig. 3. Sketch of a possible entanglement configuration

Table 3. Parameter of LDPE and HDPE-samples

Δt = 20 seconds

Sample	T/K	$y(T)$	λ_m	a
LDPE	293	45	11	0
	373	—	11	0.24
	388	—	11	0.44
HDPE2	393	—	8.8	0.34
M_w = 4000				
	413	—	8.8	0.58
HDPE1	393	—	9.3	0.23
M_w = 12000				
	413	—	9.3	0.54

segments, their transport along chain segments and across loops should not be much different. Passing through an entanglement, the direction of momentum must be flipped over, as occurs at a free chain end; this should yield about the same fluctuation in both cases. Accordingly, fluctuation of entanglements should be larger than the one of chemical junctions. This is the reason why it is possible to identify the existence of entanglements from a description of stress- or pressure-strain curves with the aid of the van der Waals equation of state.

Pressure-strain measurements carried out for high-density polyethylene are depicted in Figs. 4 and 5. The pressure-strain curves can also be

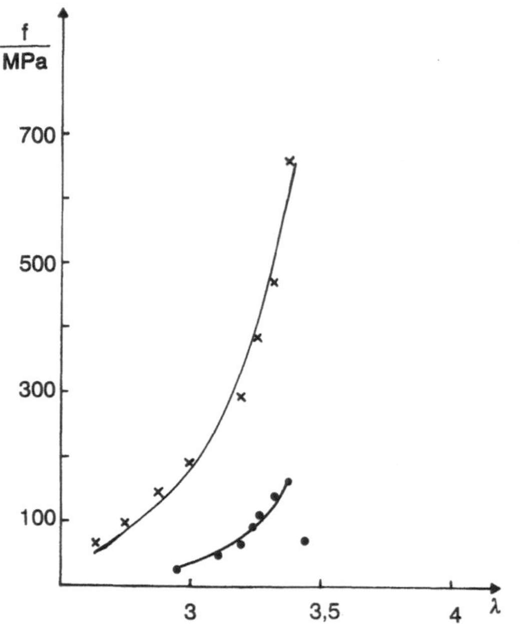

Fig. 4. Pressure against strain for HDPE1 at T = 413 K (●) and T = 393 K (*). The deformation was carried out within 20 s. The solid lines are theoretical (Eq. (2)). Parameters used are listed in Table 3

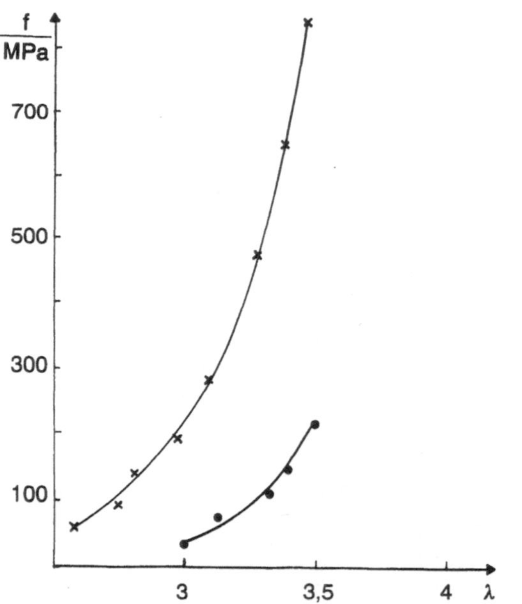

Fig. 5. Pressure against strain for HDPE2 at T = 413 K (●) and T = 393 K (*). The deformation was carried out within 20 s. The solid lines are theoretical (Eq. (2)). Parameters used are listed in Table 3

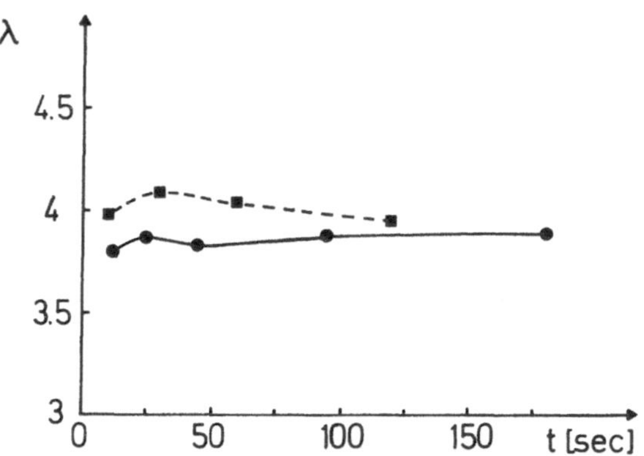

Fig. 6. Maximum strain λ of homogeneous melts under a load of 80 kN against the deformation time: ■ HDPE1; ● HDPE2

described with the aid of a quasi-permanent entanglement network. The density of junctions is identical with the one deduced from the plateau modulus (G_e = 2.2 MPa [55]; (Table 3). The network remains stable up to largest strains, without showing any identifiable dependence on the molecular weight and on the strain rate (Fig. 6). This is in accord with rheological measurements [1, 2]. What is surprising is that the density of entanglements in HDPE exceeds the one in LDPE. This is not straightforwardly understandable in terms of a molecular model [2].

The interaction parameter is, in HDPE melts, of the same magnitude of order as in LDPE (Table 3), suggesting that entanglements are at least identical in both cases.

"Non-equilibrium entanglement network"

Under adiabatic conditions the orientational entropy of a semicrystalline HDPE-sample stretched to λ = 4 should be so large as to raise the temperature for $\Delta T > 40$—$80°$. Hence, when pressing a sample at a high enough temperature, melting of all the crystals is enforced. Whatever the prehistory of deformation, the mechanical response should be determined by the momentary configuration of quasi-permanent entanglements.

"Non-equilibrium entanglement networks" are expected to be different in HDPE or LDPE. The mean distance of entanglements $\langle d \rangle$ in the homogeneous melt should be given by

$$\langle d \rangle = l_0 \lambda_m , \quad (24)$$

where l_0 is the length of the stretching-invariant unit. For polyethylene with $l_0 = 0.128$ nm and $\lambda_m \approx 10$, this length is obtained as $\langle d \rangle \approx 1.28$ nm. Because the crystals' dimensions range between 5—40 nm, entanglements as eutectic units must be dramatically redistributed during crystallization. When entanglements are squeezed into longitudinal defect layers of crystals or into amorphous regions (see Fig. 7) their final configuration is determined by the colloid structure. This leads to configurations which are different in HDPE and LDPE. The mean thickness of crystals in unstrained samples of HPDE is $\langle y_p \rangle \approx 30$—40 nm while the crystals in LDPE are very much smaller, $\langle y_p \rangle \approx 6$—8 nm. If all crystals melt during the stretch at a large enough deformation rate, entanglements cannot establish their equilibrium distribution configuration. It is very likely that non-equilibrium configurations of entanglements in HDPE and in LDPE are different.

Pressure-strain curves of LDPE and HDPE, pressed at temperatures as indicated, are shown in the plots of Figs. 2, 4 and 5. They can be fitted if one assumes that the system was transferred by intrinsic heating into the melt state. This is due to the finding that the bending of the pressure-strain curves of a crystal network differs markedly from the one of a permanent network. One learns from the data collected in Table 3 that the density of quasi-permanent entanglements is identical to the one obtained when pressing a homogeneous melt. Hence, during crystallization, disentangling of chains cannot be enforced; this is in principal accord with Fischer's freezing model [56].

It should be stressed that the true density of the quasi-permanent entanglements is deduced from the fit of the stress-strain curves, despite having the entanglements not distributed homogeneously. Because strain energy is equipartitioned among chains, even substantial differences in the spatial distribution should not to affect the modulus by more than 10—20%. The mean maximum strain parameter seems weakly to depend only on the chain-length-distribution [32, 33]. It is for this reason that details about the structure of a network cannot easily be determined by discussing the modulus.

In this situation, it is significant that the interaction parameter of LDPE and HDPE falls down to $0.2 \leqslant a \leqslant 0.35$ (see Table 3). Hence, fluctuations of

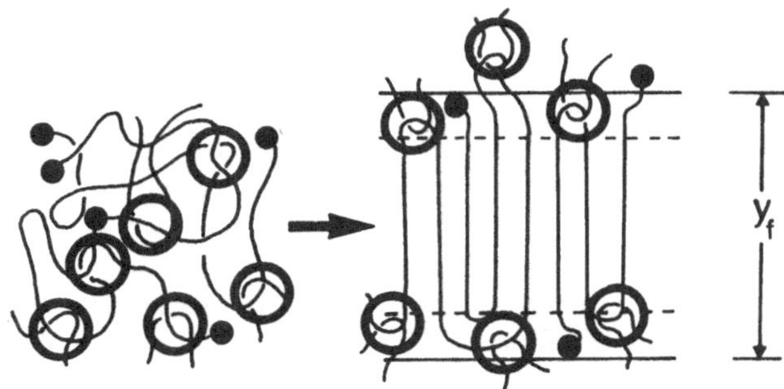

Fig. 7. Sketch of entanglement configurations in the melt (lefthand side) and within a semicrystalline system (righhand side)

entanglements in the "non-equilibrium entanglement network" seem to display values which correspond to permanent junctions with a functionality of 3 to 4 (see Table 4). This might be a consequence of having entanglements clustered. It could also be that their conformation is different than the one in Fig. 3, e.g., by having the chain-segments heavily twisted.

"Non-equilibrium entanglement networks" obtained during deformation at elevated temperatures differ from the one existing in the homogeneous melt.

Deformation of semicrystalline LDPE

Let us now ask the question whether entanglements become reorganized during deformation of semicrystalline samples. We consider it as granted that entanglements cannot be annihilated. This hypothesis is supported by a successful description of pressure-strain experiments of preoriented HDPE-samples [3].

With the aid of the equations derived, a solid description of non-isothermal pressure-strain experiments of LDPE should be possible. Whatever the prehistory of deformation [21], the effective network is always determined by λ and T; this finding facilitates our discussion. When the sample is heated during deformation, selective melting of crystals has to occur. It is the actual temperature which matters, so that a representation of the pressure-strain curves needs the correct adjustment of the parameter $y(T, \lambda)$, the thickness of smallest stable crystals. If orientation entropy effects within the rubbery regions are small, the momentary temperature T should nearly be equal to the one in the unstrained sample.

From the plot in Fig. 8, it is to be seen that the pressure-strain curve of LDPE at roomtemperature is fairly well described with the aid of Eq. (19). The parameters used are listed in Tables 1 and 3. The primary structure parameters (M_u, a, λ_m) are identical with the ones used for describing simple extension stress-strain experiments [21]. $y(T)$ must be assigned to the invariant value of 45 CH_2-units. Hence, within the accessible range of strains the sample should have been heated to a temperature of about T = 355 K. What is not straightforwardly understanable is how the temperature is raised so much in the range of relatively small strains. At T = 355 K the fractions of extended CH_2-sequence mixed crystals with a thickness of $y(T)^* < y(T)$ are molten. The degree of crystallinity is reduced (see Table 5). Because the fit depends very sensitively on the size of $y(T)$, the temperature deduced must be taken as a reliable result.

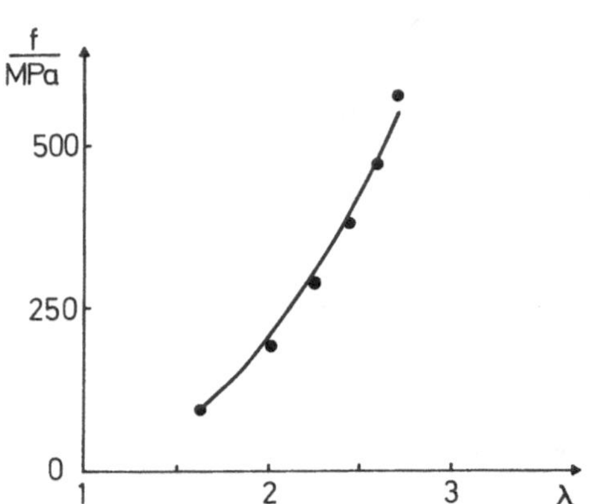

Fig. 8. Pressure against strain for HDPE1 at T = 293 K. The deformation was carried out within 20 s. The solid lines are theoretical (Eq. (19)). Parameters used are listed in Table 3

The above discussion proves that the thermodynamics of eutectoid copolymers applies during deformation. Branchings are excluded from crystal cores. The thickness distribution of the IMPs is regulated by CH_2-sequence length-distribution. The parameter a is invariant. Hence, the global mechanism of deformationn in fact, display, self-similarity. The deformation mechanism in the mode of simple extension is the same as for equibiaxial deformation.

Table 5. Parameters of the primary colloid structure of LDPE

T/K	299	313	334	353	363
w_p	0.78	0.73	0.62	0.57	0.48
$\langle y_p \rangle$	56	60	69	73	81
$y(T)$	27	31	40	44	52

The data are calculated with the aid of the parameters listed in Table 1.

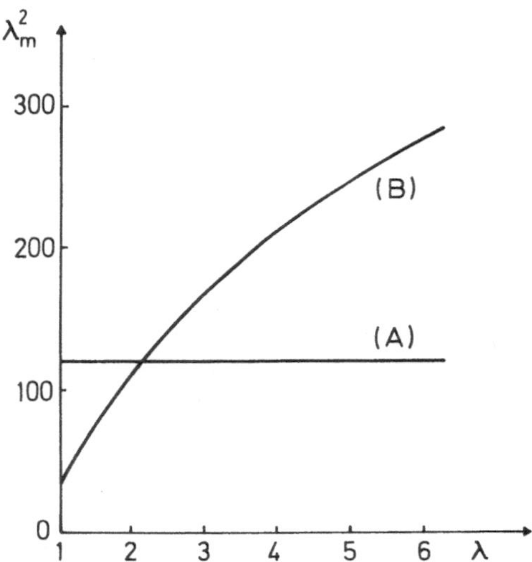

Fig. 9. Effective chain length against strain: (A) homogeneous melt and (B) semicrystalline LDPE

From the above results the hypothesis is defended that mechanically and thermodynamically equivalent subsystems of deformation should exist, the network of which defines global cooperation in a universal manner.

If no disentangling occurs it is very puzzling that the effective chain-length of the crystal-cluster network grows with strain (Fig. 9). The effective chain length exceeds the one observed in the homogeneous melt. This must imply a very sophisticated redistribution of entanglements so that the density of effective junctions is reduced (see Eq. (21)) without annihilating any entanglement. One could suggest that entanglements aggregate so as to reduce the number and increase the length of active chains. This entanglement "condensation" might also influence the ways in which the final colloid structure is developed; in simple extension, as a fibrillar superstructure. The model makes it easy to understand that density and distribution of entanglements is rapidly recovered when the system is brought into the melt. A verification of this model must be left to future experiments.

Final remarks

It has long been known that in the range of the plateau-modulus entanglements operate as quasipermanent junctions. This apparently holds true up to the largest strains achieved here. Yet, at much larger extension, the situation becomes more complicated due to finding a strain-dependent relaxation modulus [1, 2, 57—64]. According to our results, this might also imply that the entanglements configuration is changed. "Condensation of entanglements" might come into play, similar to that assumed to occur during deformation of semicrystalline samples.

This result implies that the stress-strain curve of melts should much more be bent against the λ-axis, as for a permanent network with the same density of junctions. Due to the large interaction parameter, mechanical stability should be substantially reduced [31]. This interesting aspect (which is typical for the van der Waals characterization of networks) may aid a better understanding of limitations in the mechanic stability of polymer melts.

Indeed, an interesting result of this paper is that entanglements can be identified by their large fluctuations, in comparison to what is observed for chemical crosslinks. Entanglements are possibly represented by such molecular configurations as crossing loops.

References

1. Meißner J (1981) Rheology of Polymer Melts, GdCH-Kurs, ETH Zürich
2. Laun HM (1987) Progr Coll Polym Sci 75:11
3. Kilian HG, Schrodi W, Ania F, Bayer RK, Baltà Calleja FJ, Coll Polym Sci, in press
4. Kimmich R, Schnur G, Köpf M (1988) Progr NMR Spectroscopy 20:385
5. Kimmich R, Köpf M (1989) Progr Coll Polym Sci 80:8
6. Graessley W (1982) Adv Polym Sci 47:325
7. Oppermann W, Rehage G (1981) Coll Polym Sci 259:1177
8. Dossin LM, Graessley W (1979) Rheol Acta 18:123
9. Erman B, Flory PJ (1982) Macromolecules 15:1800
10. Gottlieb M, Macosco CW, Benjamin GS, Meyers KO, Merril EW (1981) Macromolecules 14:1039
11. Gleim W, Oppermann W, Rehage G (1966) Makromol Chem 187:1273
12. Quesnel JP, Mark JE (1984) Adv Polym Sci 65:137
13. Langley NR (1968) Macromolecules 1:348
14. Capaccio G, Gibson AG, Ward IM (1979) Ultrahigh Modulus Polymers, Ed Cifferi, Ward IM, Applied Science, London
15. Smith P, Lemstra PJ, Kalb B, Pennings AJ (1979) Polym Bull 1:733
16. Barham PJ, Keller A (1985) J Mater Sci 20:2281
17. Pfeiffer DG, Kim W (1986) Lundberg RD Polymer 27:493
18. Zachmann HG (1981) Kautschuk-Gummi-Kunststoffe 34:99
19. Hoffmann M (1972) Makromol Chem 153:99
20. Hoffmann M (1972) Koll Z Z Polym 250:197
21. Treloar LRG (1940) Trans Faraday Soc 36:538
22. Hoffmann M (1977) Angewan Chem 89:773
23. Müller R, Zachmann HG (1980) Coll Polym Sci 258:753
24. Holl B, Heise B, Kilian HG (1983) Coll Polym Sci 261:978
25. Sawodny M, Asbach GI, Kilian HG (1990) Polymer 31:1859
26. Peterlin A (1971) J Mater Sci 6:490
27. Mayer J, Schrodi W, Heise B, Kilian HG (1990) Acta Polym 7:363—370
28. Peterlin A (1979) Ultrahigh modulus polymers, Ed Cifferi A, Ward IM, Appl Sci Publishers LTD, London
29. Bowdon PB, Young RJ (1974) J Mat Sci 9:2034
30. Ward IM (1975) Structure and Properties of Oriented Polymers, Appl Sci Publ LTD, London
31. Heise B, Kilian HG, Pietralla M (1977) Progr Polym Sci 62:16
32. Kilian HG, Baltà Calleja FJ (1988) Coll Polym Sci 266:29
33. Gent AN, Jeong J (1986) Polym Ing Sci 26:285
34. Van Hutten PF, Koning CE, Pennings A (1985) J Mat Sci 20:1556
35. Segula R, Rietsch F (1984) Eur Polym J 20:765
36. Ward IM (1985) Adv Polym Sci 66:81
37. Kilian HG (1981) Polymer 22:209
38. Kilian HG (1989) Macromol Chem, Macromol Symp 30:169
39. Kilian HG, Macromol Chem, Suppl in press
40. Kilian HG, Enderle HF, Unseld K (1986) Coll Polym Sci 264:866
41. Enderle HF, Kilian HG (1987) Progr Coll Polym Sci 75:55
42. Zrinyi M, Kilian HG, Horkay F (1989) Coll Polym Sci 267:311
43. Ambacher H, Enderle HF, Kilian HG, Sauter A (1989) Progr Coll Polym Sci 80:209
44. Kilian HG, Rosenberger B, Asbach GI, Wilke W, Rodrigéz (1988) Makromol Chem 189:2627
45. Kilian HG (1986) Prog Coll Polym Sci 8:689
46. Kilian HG (1988) Progr Coll Polym Sci 78:161
47. Asbach GI, Bodor G, Kilian HG (1989) Coll Polym Sci 267:976

48. Holl B, Kilian HG, Schenk H (1990) Coll Polym Sci 288:205
49. Büche F (1960) J Appl Polym Sci 4:107
50. Büche F (1961) J Appl Polym Sci 5:271
51. Kilian HG (1986) Kautschuk Gummi Kunststoffe 8:689
52. Pollak P, Asbach GI, Kilian HG, Coll Polym Sci, in press
53. Bovey FA, Winslow FH (1979) "Macromolecules", Academic Press, New York
54. Ambacher H, Enderle HF, Kilian HG, Sauter A (1989) Progr Coll Polym Sci 80:209
55. Schilling H, Pechhold W (1970) Acustica 22:244
56. Fischer EW (1978) Pure Appl Chem 50:1319
57. Lodge AS (1964) Elastic liquids, Academic press, New York
58. Wagner MH (1976) Rheol Acta 15:136
59. Wagner MH (1979) Rheol Acta 18:33
60. Laun HM (1981) Coll Polym Sci 259:97
61. Wagner MH, Stephenson SE (1979) J Rheol 23:489
62. Doi M, Edwards SF (1978) J Chem Soc Faraday Trans II 74:1818
63. de Gennes PG (1976) Macromolecules 9:587, 594
64. Klein J (1978) Macromolecules 11:852
65. Gubler MG, Kocacs AJ (1959) J Polym Sci 34:551

Received January 14, 1991
accepted February 8, 1991

Authors' address:

Prof. Dr. H. G. Kilian
Abteilung für Experimentelle Physik
Universität Ulm
Albert-Einstein-Allee 11
7900 Ulm, FRG

Inhomogeneous distributions of polarization in polyvinylidene fluoride mono- and multilayer films studied by the laser intensity modulation method

M. Wübbenhorst*) and P. Wünsche

Department of Physics, University of Leipzig, FRG
*) Present address: Department of Polymer Technology, Faculty of Chemical Technology and Materials Science, Delft University of Technology, Delft, The Netherlands

Abstract: The versatility of the laser intensity modulation method (LIMM) will be discussed concerning the interpretation of polarization distributions in multilayer films. Our LIMM equipment has been used to study monolayers, a bimorphic, and a four-layer film of PVDF. The growth of the polarization is measured in dependence on temperature for monolayer films. The poling dynamics is governed by thermally stimulated phase transitions and can be interpreted by an Arrhenius model. The LIMM capabilities and limitations are discussed for multilayer samples, including interfacial space charge effects between polarized and nonpolarized layers.

Key words: Distribution of polarization; multilayer films; polyvinylidene fluoride (PVDF)

1. Introduction

The discovery of the unique pyroelectric and piezoelectric properties of polyvinylidene fluoride (PVDF) [1, 2] started a large number of examinations on PVDF about 20 years ago, especially to understand the fundamental mechanisms of its interesting electrical properties. The essential influence on the macroscopic properties of PVDF of film preparation, stretching and poling under high electric DC-fields has been shown by using conventional methods like X-ray diffraction, IR-spectroscopy or the direct measurement of piezo- and pyroelectric coefficients. Unfortunately, all of these investigations yield results which are averaged across the polymer film. However Phelan et al. [3] demonstrated in 1974 the presence of an inhomogeneous polarization across the film by using frequency-dependent pyroelectric measurements.

These and results discussed by Marcus [4] have stimulated the development of various experimental methods to determine polarization and space charge distributions in polymer films, based either on acoustical (pressure pulse [5—7] or pressure steps [8—10]) or thermal excitation of the sample layer [11—14] were the acoustical methods are connected with both better spatial resolution of the results and more extensive experimental requirements. Nevertheless, new developments in computational and experimental techniques enhanced the spatial resolution of calculated results obtained by using thermal waves and make them more attractive [15—18].

The purpose of this paper is the presentation of some aspects of a high resolution thermal wave method, the laser intensity modulation method (LIMM) developed by Lang and Das-Gupta [13], including the heat-transfer problem for multilayers, the discussion of the pyroelectric response together with the interpretation of some experimental results using our LIMM equipment.

2. Theory

LIMM is based on exposure of the surface of a dielectric film by a periodically modulated heat flux, usually producted by a laser beam. The resulting variation of the surface temperature creates thermal waves that propagate into the sample volume and produce a definite local and time-dependent temperature variation. A pyroelectric current can be measured; it is caused by the interaction of the varying temperature with a polarization distribution $P(x)$ or space charge distribution $\rho(x)$. This $P(x)$- or $\rho(x)$-distribution can be determined from the relation between the modulation frequency $f = \omega/2\pi$ and the pyroelectric current amplitude $I(\omega)$ using a suitable deconvolution algorithm. Therefore the theoretical analysis of LIMM consists of three main parts:

I) the special heat conduction problem;
II) the interaction between temperature profile and the polar properties of the dielectrics (pyroelectric effect); and
III) the deconvolution technique.

2.1. *Heat transfer problem*

2.1.1. *Homogeneous layer*

We consider a planar sample of thickness L and assume L to be small compared to the lateral dimensions. So the relevant heat conduction equation has the following one-dimensional form

$$\rho_m c \frac{\partial T}{\partial t} = k \frac{\partial^2 T}{\partial x^2} , \tag{1}$$

where t is the time, T the temperature difference from the ambient temperature T_0, and x the spatial variable across the sample. The thermal properties of the sample are given by the density ρ_m, the specific heat c, and the thermal conductivity k, which are related to the thermal diffusion constant $K = k/\rho_m c$.

The heat transfer between the sample and its surroundings is assumed to consist of a time-dependent heat flux $q(L, t)$ at $x = L$ with

$$q(L, t) = q_0 \eta (1 + e^{i\omega t}) , \tag{2}$$

where q_0 is the laser amplitude and η the absorptivity, and a heat loss at $x = 0$ and $x = L$ due to heat conduction, characterized by the heat transfer coefficients h_0 and h_L, respectively. In our experimental equipment, freestanding polymeric layers are used and thus the parameters h_0 and h_L are equal. According to Lang and Das-Gupta [14] the temperature evolution can be separated into a transient T_u and a periodical T_ω component:

$$T = T_u + T_\omega . \tag{3}$$

By substitution of Eq. (3) into Eq. (1), we get an expression representing the transient and the periodical component of the heat-conduction problem. By separation of the variables each problem can be treated separately. This yields both the well-known plane-wall transient solution $T_u(x, t)$ [19] and the periodical solution $T_\omega(x, t)$ [14]

$$T_\omega(x, t) = \frac{(1 - i) q_0 \eta \cosh[D(1 + i)x]}{2kD \sinh[D(1 + i)L]} e^{i\omega t} , \tag{4}$$

where $D = (\omega/2K)^{1/2}$. The term $\lambda = 1/D$ is usually called the thermal diffusion length at which a thermal wave of modulation frequency $\omega/2\pi$ is attenuated to $1/e$ of the initial amplitude.

The transient component, representing the averaged sample temperature at $t \to \infty$, yields a parabolic temperature profile. Hence, we can neglect the transient component in the following because only temporal changes of the temperature contribute to a periodic pyroelectric response, and we get from Eq. (4)

$$\frac{\partial T(x, \omega)}{\partial t} = (1 + i)\omega \frac{q_0 \eta}{2kD} \frac{\cosh[D(1 + i)x]}{\sinh[D(1 + i)L]} e^{i\omega t}. \tag{5}$$

Equation (5) describes the spatial and temporal development of the temperature variation as a function of the modulation frequency. This relation allows the exact prediction of the thermal wave behavior through the sample if we know the constants q_0, η, and the sample parameters L, K and k. This is demonstrated in Fig. 1 which shows the real part of the complex amplitude $Re\{\partial T(x, \omega)/\partial t\}$ as a function of the frequency according to Eq. (5).

All curves are normalized to 1 at the position $x = L$. The thermal diffusion length can be varied from $\lambda \approx L/10$ to $\lambda \approx L$ by variation of the frequency f between 10 Hz and 10 kHz, respectively (Fig. 1).

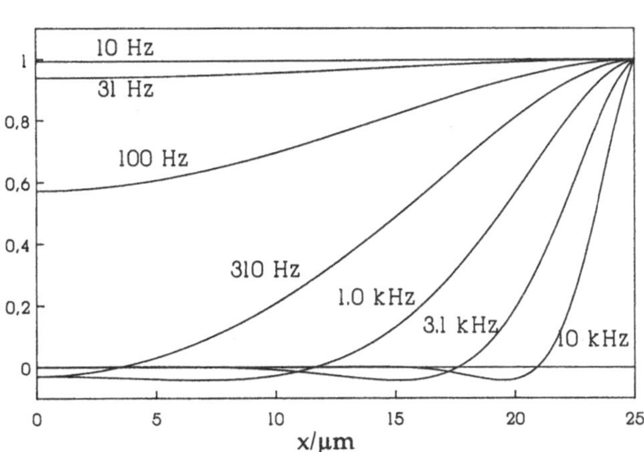

Fig. 1. Periodic component of the temperature $T_\omega(t)$ calculated for a sample with a heat flux defined by Eq. (2) and equal heat losses at both surfaces

2.1.2. *Multilayer problems*

From the point of view of heat transfer a multilayer consists of n adjacent single layers, each characterized by a local temperature distribution $T_i(x)$, thermal diffusivity K_i, and thickness L_i (Fig. 2).

Hence, the general multilayer problem can be represented by a system of n coupled differential equations

$$\frac{\partial T_i}{\partial t} = K_i \frac{\partial^2 T_i}{\partial x^2} , \tag{6}$$

$$(1 \leqslant i \leqslant n , \quad L_{i-1} \leqslant x \leqslant L_i) ,$$

each describing the heat conducting process within the ith layer. $2(n-1)$ boundary conditions are imposed due to the continuity of the temperature and the heat flux at the $n-1$ interfaces

$$T_i(L_i) = T_{i+1}(L_i) , \quad (1 \leqslant i \leqslant n) , \tag{7}$$

and

$$K_i \frac{\partial T_i}{\partial x} = K_{i+1} \frac{\partial T_{i+1}}{\partial x} \Bigg|_{L_i} , \quad (1 \leqslant i \leqslant n-1) , \tag{8}$$

as well as two boundary conditions for the outer surfaces at $x = 0$ and $x = L_n$

$$K_1 \frac{\partial T_i}{\partial x} = h(T_1 - T_u) \mid_{x=0} , \tag{9}$$

and

$$K_n \frac{\partial T_n}{\partial x} = \eta q_0(1+\exp(i\omega t)) - h(T_n - T_u)_{x=L_n} . \tag{10}$$

We additionally use the n start conditions:

$$T_i(t = 0) = T_u , \tag{11}$$

assuming that the sample initially is at ambient temperature.

Three ways are proposed to obtain the solution $\{T_i(x,t)\}_1^n$. The exact analytical calculation of $T_i(x,t)$ can be carried out by solving the set of differential equations (Eqs. (6—11)) by a Laplace transformation, see for instance [19].

The temperature profile can be approximated by a constant temperature gradient $\Delta T_i/\Delta L_i$ within each layer if the individual layers are "thermally thin". This condition mostly requires a partition of the original macroscopic layers into smaller sublayers. The advantage of the finite difference method compared to the analytical one is the conversion of the differential equations into linear difference equations [19]. Due to its numerical principle the finite difference method can be universally used, however, the convergence and numerical errors must be carefully considered. To explain the

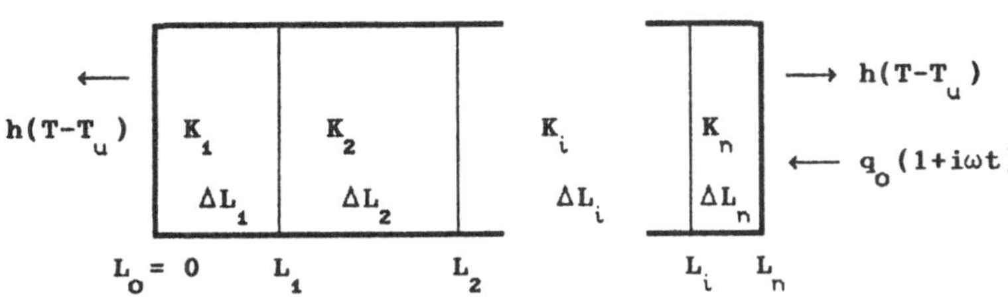

Fig. 2. Thermal model of a multilayer system

third method we introduce new spatial variables $d\bar{x}_i = dx \cdot K_i^{-1/2}$ so that the heat conduction Eq. (8) becomes

$$\frac{\partial T}{\partial t} = \frac{\partial^2 T}{\partial \bar{x}_i^2} . \tag{12}$$

Corresponding to the multilayer problem (Fig. 2), further new thickness parameters will be defined as

$$\bar{L}_i = L_i K_i^{-1/2} , \tag{13}$$

and

$$\bar{L} = \sum_{i=1}^{n} \bar{L}_i = \sum_{i=1}^{n} L_i K_i^{-1/2} . \tag{14}$$

The temperature distribution within the multilayer can be directly obtained for all spatial positions $0 \leqslant \bar{x} \leqslant \bar{L}$ by substituting the variables $\bar{x}_i$ and $\bar{L}_i$ in Eq. (5). In a final procedure the calculated $T(\bar{x}, t)$ function must be transformed into the x-representation, e.g., by a numerical integration of the relation $dx = d\bar{x} \cdot K(\bar{x})^{1/2}$.

2.2. *Pyroelectric response*

According to Collins [20] both a spontaneous polarization $P(x)$ and a space charge distribution $\rho(x)$ can contribute to the frequency-dependent pyroelectric response. ΔV_0 is the total change of the potential difference across the dielectric sample, characterized by the thickness L and the permittivity ε. So the potential difference ΔV_ρ caused only by space charges is given by [21]

$$\Delta V_\rho = \frac{1}{\varepsilon \varepsilon_0} (a_L - a_\varepsilon) \int_0^L \rho(x) \int_0^x T(\xi, t) d\xi dx . \tag{15}$$

Here, $a_L = dx/LdT$ is the thermal expansion coefficient across the sample and $a_\varepsilon = d\varepsilon/\varepsilon dT$ is the temperature coefficient of the permittivity. The second contribution to ΔV_0, caused by a permanent polarization, can be simply derived from the capacity of a capacitor containing a polarized dielectric

$$V_p = \frac{Q}{C} = \frac{PL}{\varepsilon \varepsilon_0} . \tag{16}$$

The total time derivation of Eq. (16) results in

$$\Delta V_p = \frac{PL}{\varepsilon \varepsilon_0} \left(\frac{1}{L} \frac{dx}{dT} - \frac{1}{\varepsilon} \frac{d\varepsilon}{dT} + \frac{1}{P} \frac{dP}{dT} \right) \Delta T. \tag{17}$$

The first two terms in Eq. (17) are related to a_L and a_ε, respectively. The third term $a_P = dP/PdT$ represents the pure dipolar pyroelectric effect without dimensional changes. The last component is normally attributed to a reversal change in crystallinity [22] and the temperature influence on the dipole amplitude across the sample [20]. Introducing in Eq. (16) an inhomogeneous temperature profile $T(x, t)$, we obtain

$$\Delta V_p = \frac{1}{\varepsilon \varepsilon_0} (a_p + a_L - a_\varepsilon) \int_0^L P(x) T(x, t) dx . \tag{18}$$

The alternating closed-circuit pyroelectric current $I_p(\omega)$ can be obtained by combination of Eq. (15) and Eq. (18) and the subsequent time differentiation gives

$$I_p(\omega) = \frac{A}{L} \left\{ (a_p + a_L - a_\varepsilon) \int_0^L P(x) \frac{\partial T}{\partial t} (x, \omega) dx + (a_L - a_\varepsilon) \int_0^L \rho(x) \cdot \int_0^x \frac{\partial T}{\partial t} (\xi, \omega) d\xi dx \right\} . \tag{19}$$

Integration by parts allows us to rewrite Eq. (19) as

$$I_p(\omega) = \frac{A}{L} \int_0^L \left[- (a_p + a_L - a_\varepsilon) \frac{dP(x)}{dx} + (a_L - a_\varepsilon) \rho(x) \right] \int_0^x \frac{\partial T}{\partial t} (\xi, \omega) d\xi dx . \tag{20}$$

There are two contributions $(a_p + a_L - a_\varepsilon) dP(x)/dx$ and $(a_L - a_\varepsilon)\rho(x)$ which independently can cause the resulting measurement response. It appears, therefore, impossible to distinguish a priori between space charge and polarization contributions in the measured total signal $I_p(\omega)$. One may tentatively carry out a deconvolution, assuming either a charge or polarization function. All experimental results in this paper are discussed assuming a combined function called "effective polarization" $P_{eff}(x)$, nevertheless, the deconvoluted distribution will be interpreted depending on the physical background by polarization or space charge contributions.

2.3. Deconvolution algorithm

A set of $2 * N$ discrete measuring points $\{I_0, \omega\}_1^N$ and $\{I_L, \omega\}_1^N$ corresponding to a thermal excitation at the back and front electrode is used to determine the $P(x)$- and/or $\rho(x)$-function. One well-known method to solve the Fredholm integral equation (20) is to present the unknown polarization function $P_{eff}(x)$ as a truncated orthogonal polynomial series as described by Lang and Das-Gupta [14]. Another principle is based on the numerical integration of the discrete representation of Eq. (19). We can derive an expression for the complex amplitude of the pyroelectric current at the modulation frequency $\omega_j/2\pi$ by insertion of Eq. (5) into Eq. (18), and the replacement of the integral by a sum:

$$I_L(\omega_n) = \frac{Aq_0\eta\omega_j}{2kDL\sinh[D(1+i)L]} \cdot \Bigg\{ (1+i)(\alpha_p + \alpha_L + \alpha_\varepsilon) \sum_{m1=1}^{M1} P(x_{m1}) \cdot \cosh[D(1+i)x_{m1}]x + \frac{1}{D}(\alpha_L - \alpha_\varepsilon) \cdot \sum_{m2=1}^{M2} \rho(x_{m2}) \sinh[D(1+i)x_{m2}]\Delta x \Bigg\} . \tag{21}$$

The analogous relation for the case of thermal excitation at the back surface ($x = 0$) can be obtained by replacing the spatial variable x by $L - x$.

Note that Eq. (21) contains information about both the amplitude of the measured quantity and its phase shift relative to the phase of the input heat flux. It can be shown that the projection of the complex signal onto an arbitrary phase $\Delta\phi$ contains the full information. The only exception is the case $\Delta\phi = 90°$ where the average value of the polarization distribution is not represented by $I_p(\omega)$.

In our experiments, we have mainly measured the real part of the pyroelectric current $I_L^{re}(\omega_n)$ and $I_0^{re}(\omega_n)$. The corresponding relationship for a subsequent data processing was derived from Eq. (21):

$$I_L^{re}(\omega_n) = \frac{Aq_0\eta}{2kL} \Bigg\{ (\alpha_p + \alpha_L - \alpha_\varepsilon) \cdot \sum_{m1=1}^{M1} B^1_{n,m1} P_{m1}(x_{m1}) + (\alpha_L - \alpha_\varepsilon) \cdot \sum_{m2=1}^{M2} B^2_{n,m2} \rho_{m2}(x_{m2}) \Bigg\} , \tag{22}$$

(for $n \leqslant N$) with

$$B^1_{n,m1} = \frac{\omega_i \Delta x_{m1}}{D[1+\cosh(2DL)]} \cdot (\cos(DL)\sinh(DL)[\cos(Dx)\cosh(Dx) - \sin(Dx)\sinh(Dx)] + \sin(DL)\cosh(DL)[\cos(Dx)\cosh(Dx) + \sin(Dx)\sinh(Dx)]) , \tag{23}$$

and

$$B^2_{n,m2} = \frac{2K\Delta x_{m2}}{1+\cosh(2DL)} (\cos(DL)\sinh(DL) \cdot \cos(Dx)\sinh(Dx) + \sin(DL) \cdot \cosh(DL)\sin(Dx)\cosh(Dx)) . \tag{24}$$

The restriction $n \leqslant N$ for the indices n in Eq. (22)–(24) is inevitable if we define a common input vector for all measured data points $I_L(\omega)$ ($1 \leqslant n \leqslant N$) and $I_0(\omega)$ ($N+1 \leqslant n \leqslant 2N$) corresponding to the front and back electrode irradiation

$$I = [I_L(\omega_1) \ldots I_L(\omega_N) , \ I_0(\omega_1) \ldots I_0(\omega_N)]^T . \tag{25}$$

The coefficients $B^1_{n,m1}$ and $B^2_{n,m2}$ for $N+1 \leqslant n \leqslant 2N$ can be obtained from Eqs. (23) and (24) by an index exchange of m_1 by $M_1 - m_1 + 1$ and of m_2 by $M_2 - m_2 + 1$. Finally, the linear system of equations in matrix representation is

$$\begin{bmatrix} I \\ \vdots \\ I_{2N} \end{bmatrix} = C_1 \begin{bmatrix} B^1_{1,1} & B^1_{1,2} & \dots & B^1_{1,M1} \\ \vdots & & & \vdots \\ B^1_{2N,1} & B^1_{2N,2} & \dots & B^1_{2N,M1} \end{bmatrix} \cdot \begin{bmatrix} P_1(x_1) \\ \vdots \\ P_{M1}(x_{M1}) \end{bmatrix}^T$$

$$\oplus C_2 \begin{bmatrix} B^2_{1,1} & B^2_{1,2} & \dots & B^2_{1,M2} \\ \vdots & & & \vdots \\ B^2_{2N,1} & B^2_{2N,2} & \dots & B^2_{2N,M2} \end{bmatrix}$$

$$\cdot \begin{bmatrix} \rho_1(x_1) \\ \vdots \\ \rho_{M2}(x_{M2}) \end{bmatrix}^T , \tag{26}$$

or

$$I = \mathbb{B}F , \tag{27}$$

where the partial matrices $\mathbb{B}^1$ and $\mathbb{B}^2$ are combined into a matrix $\mathbb{B} = C_1\mathbb{B}^1 \oplus C_2\mathbb{B}^2$ and the vector F of the dimension $M_1 + M_2$ is defined by $F = [P,\rho]^T$. The constants C_1 and C_2 can be evaluated by comparison of Eq. (22) with Eq. (26). The normal solution of Eq. (27) is characterized by the pseudo inverse matrix $\mathbb{B}^+$

$$F = \mathbb{B}^+ I . \tag{28}$$

This Fredholm integral equation of the first kind leads to illconditioned normal equations which are very sensitive to numerical and experimental errors. Therefore, it is advantageous to use a least-square procedure modified by regularization [24] with the regularization coefficient γ

$$\| I - \mathbb{B}F \|^2 + \gamma \| F \|^2 \Rightarrow \text{Minimum} . \tag{29}$$

The matrix $\mathbb{B}$ can be written as $\mathbb{B} = \mathbb{U} \sum \mathbb{V}^T$, using the singular value decomposition (SVD) by Golub and Reinsch [25], where $\sum$ is defined by

$$\sum = \mathbb{U}^T \mathbb{B} \mathbb{V} = \text{diag}(\sigma_1, \dots, \sigma_l) \in \mathbb{R}^{m,n} , \tag{30}$$

with

$$l := \min\{m, n\} ,$$

and

$$\sigma_1 \geqslant \sigma_2 \geqslant \dots \geqslant \sigma_r > 0 ,$$

$$\sigma_{r+1} = \sigma_{r+2} = \dots = \sigma_l = 0 .$$

Here, $\sum$ is a diagonal matrix with r nonzero diagonal elements. The regularized solution F_γ can be finally obtained by

$$F_\gamma = \mathbb{V}[\textstyle\sum(\sum^2 + \gamma\mathbb{E})^{-1}]\mathbb{U}^T I . \tag{31}$$

This algorithm is the basis for handling the measured $I(\omega)$-data (Fig. 3).

The measured current-frequency data set consists of logarithmically spaced $I_L(\omega)$- and $I_0(\omega)$-points in the frequency range from $f_u = 10$ Hz to $f_0 = 10$ kHz (Fig. 3 (1)). The subsequent deconvolution provides the polarization distribution, assuming a certain sample thickness L and thermal diffusivity K (Fig. 3 (2)). With this polarization distribution an "ideal current-frequency function" is calculated using Eq. (22) which is compared with the measured current-frequency function (Fig. 3 (3)).

The described procedure can be repeated partly or completely to study the influence of errors in the input sample parameters, e.g., a thickness error $\Delta L/L$. This is necessary due to the complex relationship between the measured quantities and the final distribution function. Hence, a number of simulations have been made [17] to estimate the influence of errors in L and K, at different noise levels superimposed with the $I(\omega)$-data, as well as the ratio I_L/I_0 (caused by different absorptivities of both sample electrodes), and the shape of the polarization profile itself. It turns out that the most critical parameter is the regularization parameter γ, because the resulting vector F_γ strongly depends on the parameter γ for a given input vector I. The "best" value of the parameter γ cannot be derived from theory, hence a systematic parameter variation has been carried out. $\langle\Delta I\rangle = (I_i^{mea} - I_i^{calc})^2$ has been calculated (Fig. 4 (a—c)) as an objective measure to find out the optimum parameters were I_i^{mea} and I_i^{calc} are the measured and calculated values, respectively.

Figure 4 (a) demonstrates the interdependence of the thickness and thermal diffusivity. The valley depicted in Fig. 4 (a) is consistent with $K \propto L^{-1/2}$, corresponding to the heat conduction equation.

Figure 4 (b) shows a local minimum of $\langle\Delta I\rangle$ which indicates the independent influence of thermal diffusivity and the I_L/I_0 ratio on $\langle\Delta I\rangle$. Consequently, both parameters can be optimized independently with the help of the error criterion $\langle\Delta I\rangle$.

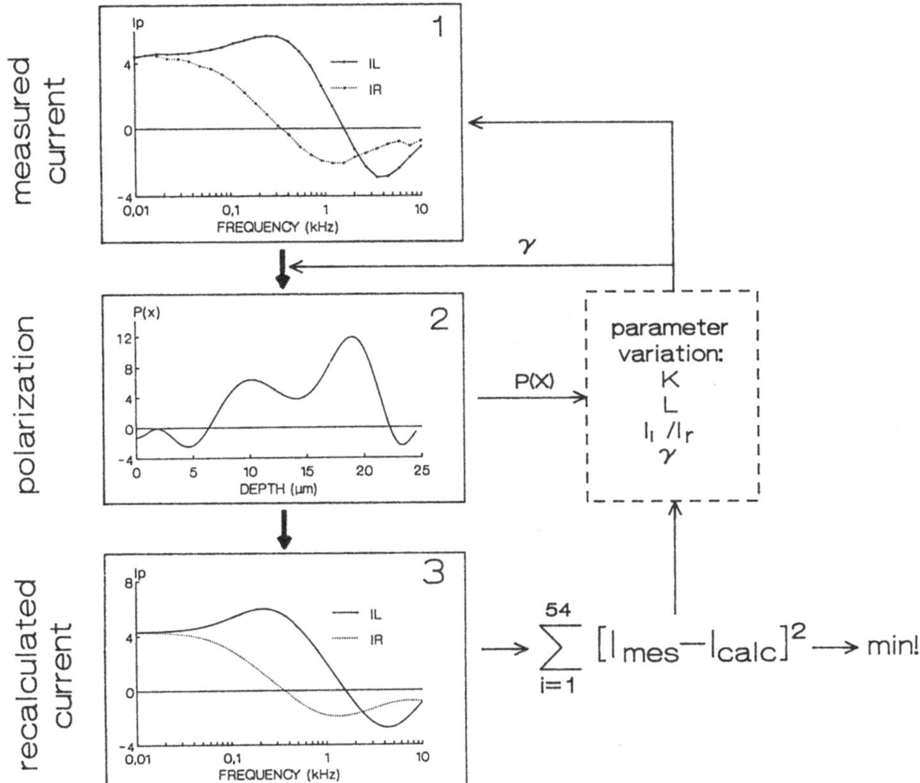

Fig. 3. Illustration of the algorithm to calculate the effective polarization $P_{eff}(x)$

Finally, the influence of the most difficult parameter γ is shown in Fig. 4 (c) combined with a variation of the assumed sample thickness. We do not find a two-dimensional local minimum as expected; we can only see a minimum for the thickness. Hence, it is necessary to define an upper limit for $\langle \Delta I \rangle$ to estimate the optimum regularization parameter γ.

3. Experimental

Figure 5 shows the schematic representation of the measurement equipment used for our LIMM measurements. A 50 mW He-Ne-laser provides a constant radiation flux which is modulated by a sinusoidally driven acousto-optical modulator (AOM).

Only the extraordinary output beam of the AOM is used because its intensity can be varied between nearly zero and about 40% of the input intensity. The beam irradiates the sample placed in a special sample chamber after passing some mirrors and lenses for guiding and focussing. Aluminum was evaporated onto both sides of the sample to obtain optically opaque electrodes and a high surface conductivity. An additional coating with black ink was applied to enhance the absorptivity. The sample itself has been clamped between two PMMA-rings with an inside diameter much larger than the pyroelectric active circular metallized area. This fact, as well as the use of very thin wires for electrical contacts guarantees the required thermal isolation expressed by a high thermal time constant τ_{th}.

The pyroelectric response is amplified by a current voltage converter which keeps to zero the electrical boundary condition (zero external field

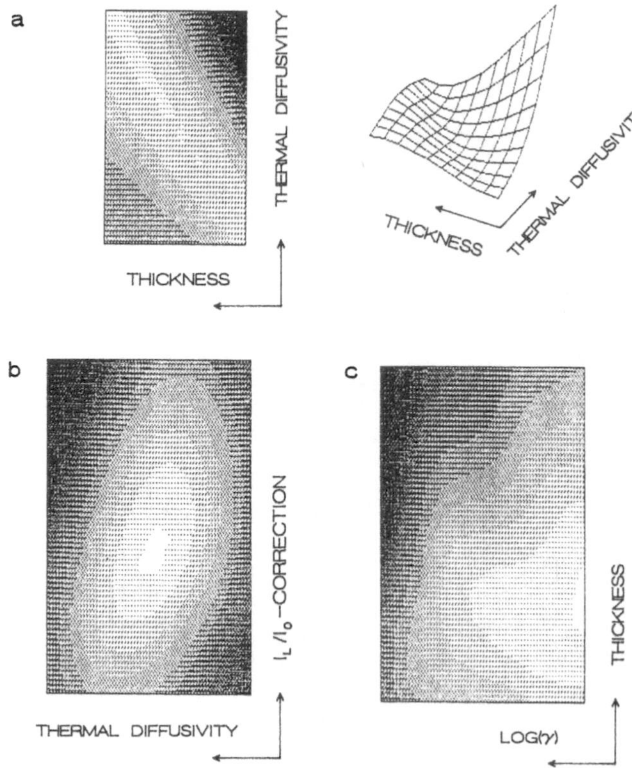

Fig. 4. Two-dimensional representation of $\sum (I_i^{mea} - I_i^{calc})^2$ for the variation of:
a) thickness and thermal diffusivity;
b) thermal diffusivity and I_L/I_0 ratio;
c) $\log(\gamma)$ and thickness

strength). A normal resistor R works as a current voltage converter. The condition $(RC_p)^{-1} \gg f_0$ must be fulfilled where C_p and f_0 denote the sample capacity and the highest modulation frequency. The application of an active current voltage converter is by all means necessary for an improved sensitivity of the electrical input circuit.

The small AC voltage is amplified by a broad band preamplifier UNIPAN 223-7 followed by a lock-in amplifier UNIPAN 233 processing the frequency range from 1 Hz to 150 kHz. Either the real or imaginary part of the amplitude of the complex pyroelectric response can be obtained depending on the phase shift $\Delta\phi$ relative to the reference signal generated by a digital controlled oscillator. The lock-in output voltage $U_p(\omega, \phi)$ will be transformed by an analog digital converter into a digital representation for further analysis. The equipment is fully controlled by a microcomputer. Use was made of a Turbo Pascal program for the data analysis, including deconvolution and graphics routines.

It should be noted that the LIMM does not provide the absolute values of the polarization and/or space charge distribution. This is due to uncertainty regarding the sample absorptivity, as well as the coefficients α_L, α_ε, and α_p. Therefore, an additional measurement of the quasistatic (averaged) pyroelectric coefficient was used to calibrate the obtained polarization profiles. We will use the pyroelectric coefficient and the polarization in the same

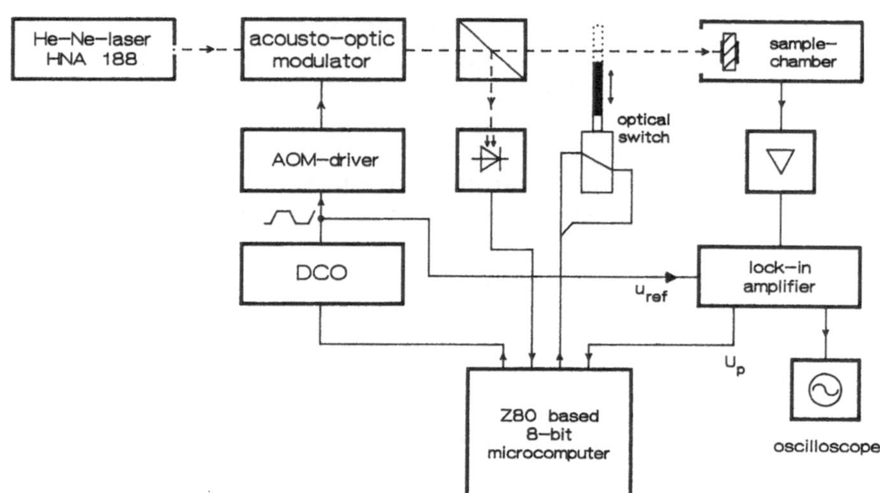

Fig. 5. Schematic representation of the experimental equipment

meaning in the following because of the approximately linear relation between them in each kind of PVDF material.

4. Results and Discussion

The above-mentioned experimental procedure has been used in an extensive study of nonhomogeneous polarization effects in films of PVDF [17, 26], PA-6, PA-11 [18] and other polymeric materials. The poling behavior of all these materials has been investigated, i.e., the formation of a permanent macroscopic polarization within an high external field under definite poling conditions. Treatment of the polymer films usually at high temperatures initiates several processes in the polymeric solid, e.g., ferroelectric switching, conformational changes, field-induced transitions between crystalline modifications, and charge injection and transport phenomena. The semicrystalline polymers mentioned above exhibit similar poling effects regarding the spatial distribution, the magnitude, and dynamics of the pyroelectric effect. The following selected results on PVDF samples can therefore often be extrapolated to other polar polymer materials.

4.1. PVDF monolayers

4.1.1. Homogeneous polarized films

The LIMM procedure was first tested by use of the so-called completely polarized PVDF films [27] which are commercially available as "piezofilms" from several companies. In detail, we investigated two PVDF-foils (made by Solvay & Cie Bruxelles) which are known to be poled by a positive corona discharge [28]. The foils have a thickness of 25 and 9 μm, respectively, and are called S25bp and S9bp. Figure 6 shows the calculated polarization profiles. Both films show a relatively constant effective polarization within the sample, and an asymmetrically shaped polarization near the surfaces. Because of the highest experimental resolution of the LIMM in the surface region, we can consider this asymmetrical polarization decay as a real effect, caused by either structural inhomogeneities or an asymmetrical electric field profile during the poling. Consequently, for this type of sample a spatial resolution of better than 0.5 μm is estimated. But it is a special property of the LIMM that the spatial resolution decreases from the surface to the central region.

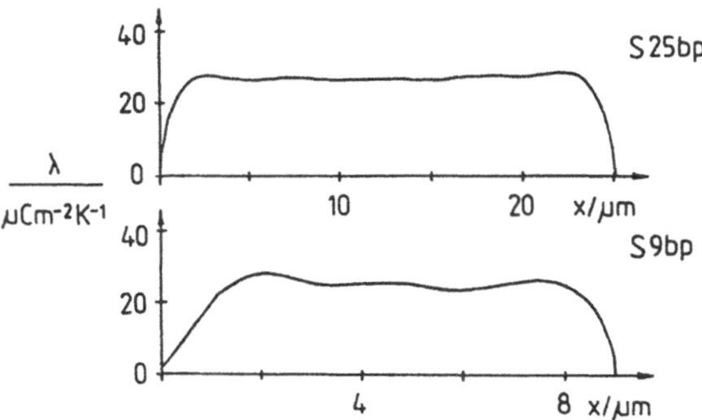

Fig. 6. Polarization profiles of two Solvay-PVDF foils with a thickness of 25 μm and 9 μm for the samples S25bp and S9bp, respectively

4.1.2. Polarization dynamics

The LIMM was applied for systematic investigations on the temporal evolution of inhomogeneous polarization profiles. These measurements were made with mainly α-phase (nonpolar crystalline form II) containing PVDF-films at relatively weak poling field strengths (Ep = 100 MVm^{-1}), where the appearance of inhomogeneous polarization distributions might be expected [27].

It is generally accepted that the presence of injected or intrinsic charges influences the formation of a permanent polarization in PVDF [8-10, 27, 29]. Eisenmenger and Haardt first described the interaction between space charges and dipolar polarization by a polarization dependent trap-model (TPP), assuming an enhanced trapping probability near polarized crystalline regions resulting in local charge compensating zones at the crystallite surfaces [8, 29]. Following this idea, Holdik and Bihler studied the temporal development of polarization profiles with in situ pressure pulse step technique (PPS) during poling at room temperature [9, 10, 27].

The present LIMM measurements have been made in order to study the influence of the poling temperature on the poling dynamics.

Sample preparation and characterization

The samples were weakly oriented, unstretched PVDF films made by Bamberg GmbH, FRG. The phase composition, consisting of both form-I (β-phase) and form-II (α-phase) fractions, was estimated by infrared spectroscopy [30] and presented a volume-ratio of 9:1 for the α phase relative to the β one. All 50-μm-thick films were metallized in vacuum with approximately 100-nm-thick aluminum electrodes on both sides (circles 4 mm in diameter). Samples were poled under controlled poling conditions E_p (poling field strength), T_p (poling temperature), and t_p (poling time). Each poling experiment was stopped after a definite poling time and the sample was investigated by LIMM in order to study the dynamics of polarization measurement procedure.

Poling dynamics at room temperature

The effective polarizations $P_{eff}(x)$ are shown in Fig.7a for a series of PVDF samples poled at room temperature during various poling times.

Figure 7a shows asymmetric polarization profiles with a peak near the positive electrode. This behavior is either caused by different charge mobilities of positive and negative charge carriers, or by asymmetric charge injection from the electrodes corresponding to the TPP model.

The maximum pyroelectric coefficient value is (35—40) μCm^{-2} K^{-1} for pure β-PVDF films, indicating a contribution of about (3.5—4.0) μCm^{-2} K^{-1} to the pyroelectricity [17] by the initial β part of the material used. In fact, the poling profiles could be interpreted up to 60 s as being only caused by ferroelectric switching of the β phase. Above this temporal interval the positive polarization peak increases slower and obtains its final state after nearly 3000 s. This "long time process" can be attributed to the field induced $\alpha \rightarrow \delta$ phase transition, which needs higher electrical field strengths. Consequently, a transition of the modification is supposed to be only restricted to the anode region.

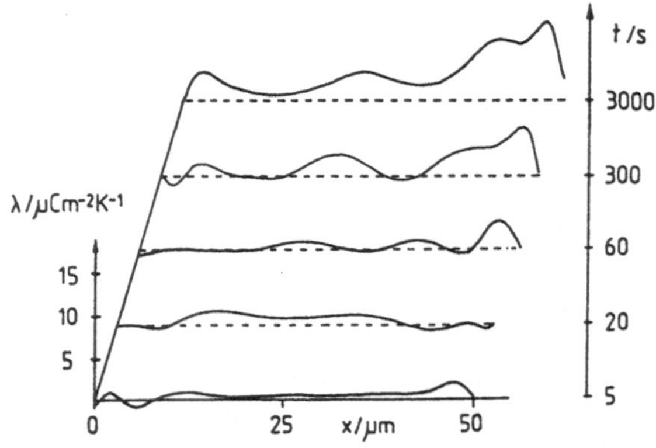

a

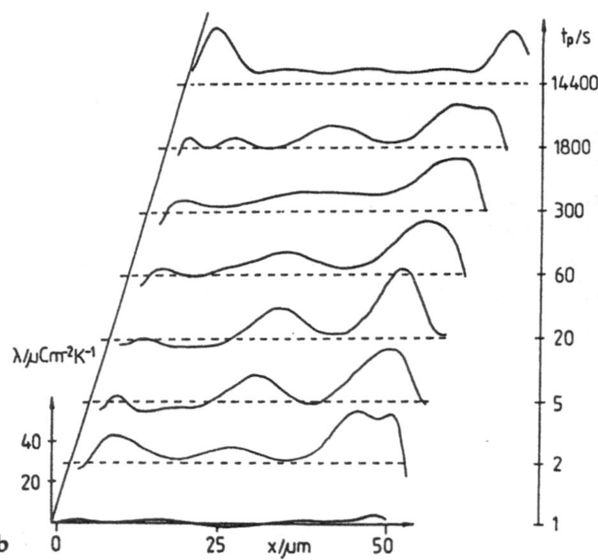

b

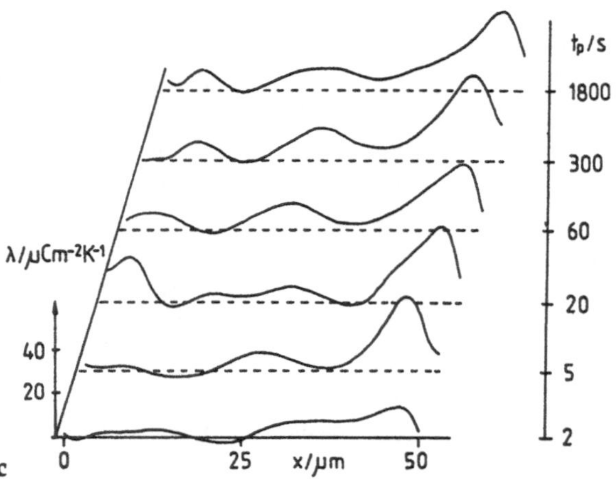

c

Fig. 7. Effective polarization $P_{eff}(x)$ of a 50-μm-thick PVDF sample in dependence on the poling time t.
Poling field strength: E_p = 1 $MVcm^{-1}$;
Poling temperatures: a) T_p = 20°C
b) T_p = 60°C
c) T_p = 80°C

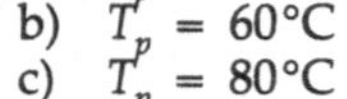

Poling dynamics of thermally poled PVDF films

Figure 7b shows the polarization profiles obtained at a poling temperature of T_p = 60°C. The results show a drastically accelerated temporal development of the polarization, as well as a general increase of maximum pyroelectric activity up to 30 μCm^{-2} K^{-1}. The steady-state situation was reached after a few seconds, indicating a very quick formation of a large fraction of polar phase. The expected increase of the β phase volume fraction could not be confirmed by using infrared spectroscopy, thus the observed enhancement of poling efficiency can only be attributed to the $\alpha \rightarrow \delta$ transformation.

Additional PVDF series, poled at T_p = 80°C, confirm this tendency and show a similar picture to that discussed above (Fig. 7c). In addition, a further increase in pyroelectric maximum values of about 10—20% indicates the appearance of high field $\alpha \rightarrow \beta$ conformational transitions.

Quantitative interpretation of poling dynamics

It is remarkable that the time for reaching the final polarization profile decreases from about 3000 s at T_p = 25°C to a few seconds already at T_p = 60°C. It was attempted to explain this immense temporal acceleration on the basis of a thermally activated process. We estimated an activation energy of $E_a \cong$ 2 eV, assuming an Arrhenius type behavior:

$$t_p(T) = t_0 \exp\left(-\frac{Ea}{kT}\right), \quad (32)$$

where t_0 and E_a are the preexponential factor and the activation energy, respectively. Both the dynamics of the various dipolar switching processes itself and the transport of real charges into the polarized regions could contribute to the resulting poling dynamics, including its temperature behavior in accordance with the TPP-model [27]. But the pure electrical conduction in PVDF films is usually characterized by an activation energy of about 1 eV [27, 31]. Therefore, some results from the literature has been fitted to a simple model proposed by Furukawa et al. [32]:

$$P(t) = P_0\left(1 - \exp\left[-\frac{t}{t_s}\right]\right), \quad (33)$$

in order to decide whether the conduction or the switching processes are dominant with respect to the temperature dependence of the poling behavior. Here, $P(t)$ respresents the time-dependent volume fraction of the polar phase, following the switching process induced by the electrical field characterized by an averaged switching time t_s and the critical field strength E_k:

$$t_s = t_{so} \exp\left[-\frac{E}{E_k}\right]. \quad (34)$$

A relation between the three main poling parameters E_p, t_p, and T_p can be derived by combination of Eqs. (34) and (35) with the Arrhenius approach of Eq. (32). We determined some different values of the activation energy for these elementary processes, assuming the description of the ferroelectric switching of the β phase, as well as of the field-induced modification transitions $\alpha \rightarrow \delta$ or $\alpha \rightarrow \beta$ by the same model. Hence, we obtained, e.g., E_a = (0.5—0.6) eV for the ferroelectric switching at temperatures from —60°C to 20°C [32], and E_a = 0.2 eV for the high field $\alpha \rightarrow \beta$ conformational transition at higher temperatures by using experimental data from the literature [33]. However, these results indicate that the temperature dependence of the poling dynamics is mainly influenced by the thermal activation of the charge transport. On the other hand, the high activation energy obtained for the entire poling dynamics cannot be explained by a normal superposition of the mentioned elementary phenomena, but rather by accepting a positive feedback mechanism in accordance to the TPP model.

3.2. PVDF multilayers

The following results were obtained from PVDF multilayer samples of prepolarized PVDF films S25bp and S9bp, and 25-μm-thick nonpolarized PVDF material S25bu. The films were glued to each other with a two-component epoxy resin that formed about 5-μm nonpolar interface layers between the PVDF films. Both surfaces of the obtained PVDF stacks were metallized under vacuum conditions.

Using this procedure a variety of multilayers were prepared consisting of two, three or four layers in different arrangements with respect to the polarization state and polarization direction.

The obtained films were studied for two main reasons:

- to find out the LIMM capabilities and limitations for multilayer samples, and
- to study the interfacial space charge effects between polarized and nonpolarized PVDF-films.

Combination of polarized and nonpolarized films

The theoretical polarization profile of an ideal asymmetric monomorphic layer is shown in Fig. 8a. Whereas the left part of the profile can be attributed to the polarized layer, the right part exhibits no polarization. The complete deconvolution procedure was tested with an ideal polarization distribution $P(x)$ and relevant sample parameters in order to investigate the influence of numerically caused deviations, for instance. The recalculated polarization functions are presented in Fig. 8a for two different regularization parameters γ together with the original $P(x)$ function. The tendency towards a nearly perfect reproducibility of the polarization distribution in the surface layer is easily recognized, whereas the details in the middle of the layer cannot be reproduced correctly. An analogue analysis can be seen for the combination of two polarized layers in series in Fig. 8c.

The observed negative part of the from measured data polarization profile measured data calculated

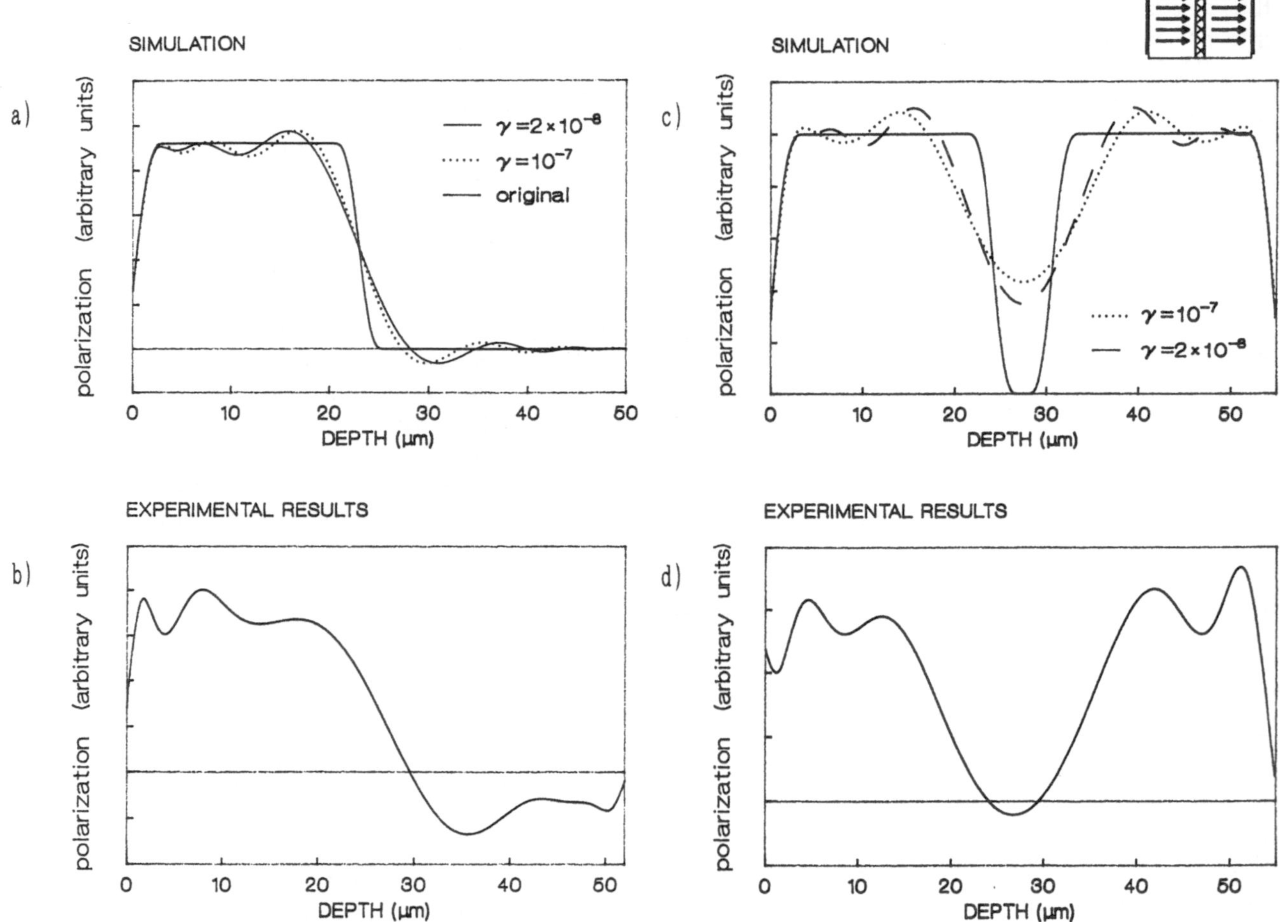

Fig. 8. Polarization profiles of bimorphic films.
One polarized and one unpolarized layer: a) model calculations; b) experimental result.
Two polarized layers in series: c) model calculations; d) experimental results

(Fig. 8b) indicates a significant pyroelectric activity in the nonpoled layer which cannot originate from a permanent polarization. Spatially distributed space charges $\rho(x)$ can also contribute to the pyroelectric signal according to Eq. (20). The surface charges σ_{PO} and σ_{PL} induced by polarization compensate the polarization-caused electrical field to zero if the polarization is constant within the sample [14].

The charge distribution becomes much more complicated in the case of multilayers. Starting with totally or partially charge-compensated polarized single layers in preparing the multilayer sample, we obtain additional layers where a part of the compensation charges will be localized. In detail, one interlayer of adhesive causes two charge layers located near the polymer-resin interface.

In order to discuss an internal field distribution proportional to the induced polarization, we made the following assumptions:

i) all polarized single layers are equally and homogeneously polarized;
ii) the polarization of each single layer is compensated by a discrete surface charge layer (also called "local compensation"), and
iii) the degree of compensation can vary from 0% (no compensation) to 100% (total compensation).

The case of completely locally compensated polarization does not yield an internal field by an induced polarization. However, the compensation charges reduce the total pyroelectric effect to the dipolar component a_p, as discussed by Collins [20]. In consequence, an incomplete charge comensation of the polarized PVDF layer corresponding to Fig. 8a should result both in a higher pyroelectricity in the polarized (left) layer, and in a negative pyroelectric activity in the unpoled (right) part of the sample. In fact, the experimentally obtained curve in Fig. 8b confirms the assumed incomplete local charge compensation.

The polarization distribution of a bilayer with two polarized PVDF films in series (Fig. 8d) exhibits a gap in the pyroelectric profile which is significantly deeper than expected (Fig. 8c). Considering the limited experimental resolution in the middle of the layer the observed minimum requires the presence of a strong negative pyroelectric activity in the central region. This effect is in agreement with the internal field distribution which corresponds to a low degree of local charge compensation (<50%). A probable explanation for this effect may be the partial neutralization of the expected double layer of the compensation charges in their own electric field within the adhesive layer.

Four-layer films

Figure 9 shows the polarization profile of a multilayer consisting of four 9-µm-thick polarized PVDF films. The direction of the polarization is given at the top of Fig. 9. The polarization profile of this four-layer film can be represented by +0+0—0—, where 0 is the adhesive layer. The polarization profile could be resolved principally by LIMM. Only the thickness of the layers is not equally reproduced due to the inhomogeneous spatial resolution, as discussed before. However, there is a remarkable difference in the pyroelectric activity between the left (positive) and the right (negative) parts of the sample. The four-layer sample can be interpreted as a sandwich of two oppositely directed bimorphic samples in analogy with the described effects at the symmetric bimorphic layer (Fig. 8b).

Both the differences in peak heights and depths of the minima seem to indicate a corresponding different degree of local charge compensation in the bimorphic layers (at about 8 µm and 32 µm). Conse-

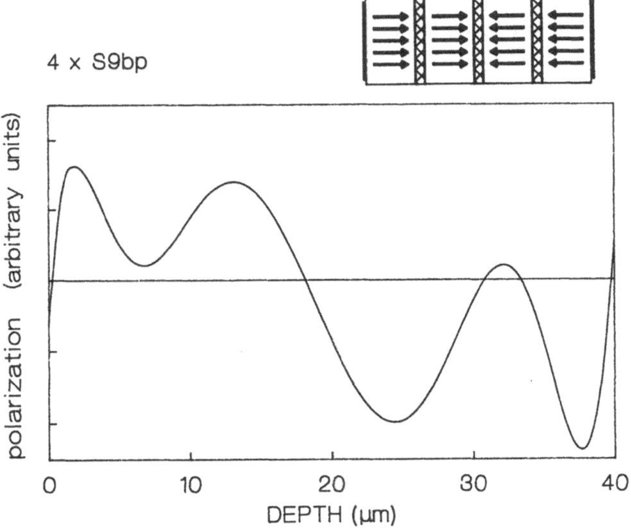

Fig. 9. Experimental polarization profile of a four-layer film

quently, the right bimorphic layer may exhibit a lower charge density at its interfaces relative to the left one. A probable reason for this difference might be variations in the electrical conductivity of these interlayers, for instance, caused by a different thickness or chemical composition of the epoxy resin.

References

1. Kawai H (1969) Jap J Appl Phys 8:975
2. Bergmann JG Jr, McFee JH, Crance GR (1971) Appl Phys Lett 18:203–205
3. Phelan R JJr, Peterson RL, Hamilton CA, Day GW (1974) Ferroelectrics 7:375–377
4. Marcus MA (1981) J Appl Phys 52:6273
5. Rozno AG, Gromov VV (1979) Soviet Technical Letters 5:266–267
6. Gerhard-Multhaupt R, Haardt M, Eisenmenger W, Sessler GM (1983) J Phys D: Appl Phys 16:2247–2256
7. Gross B, Gerhard-Multhaupt R, Berraissoul A, Sessler GM (1987) J Appl Phys 62:1429–1432
8. Haardt M (1982) Ph D thesis University of Stuttgart
9. Holdik K, Eisenmenger W (1985) In: Proc of the 5th Int Symp on Electrets, Heidelberg, pp 553–558
10. Bihler E, Holdik K, Eisenmenger W (1988) In: Proc of the 6th Int Symp on Electrets, Oxford, pp 13–17
11. Collins RE (1975) Appl Phys Lett 26:675–677
12. Lang SB, DeReggi AS, Mopsik FI, Broadhurst MG (1983) J Appl Phys 54:5598–5602
13. Lang SB, Das-Gupta DK (1984) Ferroelectrics 60:23–26
14. Lang SB, Das-Gupta DK (1986) J Appl Phys 59:2151–2160
15. Lang SB, Qing-Rui Y (1987) Ferroelectrics 74:357–386
16. Bauer S, Ploss B (1988) In: Proc of the 6th Int Symp on Electrets, Oxford, pp 28–36
17. Wübbenhorst M (1989) Ph D thesis university Leipzig
18. Thiele V, Wübbenhorst M, to be published in Acta Polymerica
19. Myers GE, Analytical Methods in Conduction Heat Transfer, (McGraw-Hill, New York, 1971)
20. Collins RE (1976) J Appl Phys 47:4804–4808
21. Collins RE (1977) Rev Sci Instr 48:83–91
22. Kepler RG, Anderson RA, Lagasse RR (1982) Phys Rev Lett 48:1274–1277
23. Kielbasinski A, Schwedlick H, Numerische Lineare Algebra, (VEB Deutscher Verlag der Wissenschaften, Berlin, 1988)
24. Tyhanov (1963) Docl Akad Nauk SSSR 153:49–52
25. Golub GH, Reinsch C (1970) Numer Math 14:403–420
26. Wübbenhorst M, Petzsche T, Ruscher C (1988) Ferroelectrics 81:373–376
27. Holdik K (1985) Ph D thesis University of Stuttgart
28. Gerliczy G, Betz R (1987) Sensors and Actuators 12:207–223
29. Eisenmenger W, Haardt M (1982) Solid State Comm 11:917
30. Holstein P (1989) private communication
31. Das-Gupta DK, Doughty (1980) J Appl Phys 51:1733–1737
32. Furukawa T, Date M, Johnson GE (1983) J Appl Phys 54:1540
33. Danz R (1982) Acta Polymerica 33:1–8

Received Februare, 21, 1991
accepted March, 22, 1991

Authors' address:

Dr. M. Wübbenhorst
Dept. of Polymer Technology
Delft University of Technology
P.O. Box 5045
2600 GA Delft, The Netherlands

Discussion

WENDORFF:
What about using blocked electrodes to distinguish between space charge effects and dipolar effects?

WÜNSCHE:
The comparison of LIMM results with and without blocking electrodes can give some hints for the relative contributions of charge trapping and dipoles, but the effect of blocked electrodes was not investigated. We have only used aluminum as electrode material up to now.

KREMER:
What limits the spatial resolution of the Laser Intensity Modulation Method (LIMM)?

WÜNSCHE:
The resolution increases in the surface region with increasing frequency, whereas the resolution remains nearly constant in the center of the film. A resolution of about 0.5 μm can be reached near the surface with frequencies up to 10 kHz.

KILIAN:
Could you comment on the influence of the structure on your results?

WÜNSCHE:
Time behavior and attainable absolute values of the polarization are strongly influenced by the content of *a*-

and β- phase in the investigated PVDF films. Additionally, the calculated polarization peaks may be caused by structural phenomena near the layer surface, but this must be investigated by additional experiments.

KREMER:
What is the degree of trapping of the mobile charge carrier considered in the evaluation of the measured currents?

WÜNSCHE:
The combination of thermally stimulated currents, with its conclusions concerning the distribution of the activation energy, and LIMM seems to open a possibility to estimate the relative contributions caused by trapped charges or dipoles, respectively.

Progress in Colloid & Polymer Science Progr Colloid Polym Sci 85:38—45 (1991)

Thin monomolecular polymer layers as low-dimensional polymer systems

T. M. Birshtein, O. V. Borisov, A. A. Mercurieva, and E. B. Zhulina

Institute of Macromolecular Compounds of the Academy of Sciences of the USSR, Leningrad, USSR

Abstract: The theory of phase transitions in monomolecular polymer layers is developed as in low-dimensional systems. Two types of planar layers are considered: the layers formed by the chains lying on the surface or by the chains grafted onto it at one end. In the former case the system is two-dimensional, whereas in the latter case the chains are effectively one-dimensional. It is shown that the character of conformational transitions in these low-dimensional systems differs markedly from that of transitions in ordinary three-dimensional systems. According to the mean-field theories, the LC-ordering in a two-dimensional system of anisotropic or anisotropically interacting particles loses its jumpwise character and becomes the second-order phase transition (with the exception of a system of particles oriented along the axes of the hexagonal lattice). In a quasi-one-dimensional system of chains grafted at one end onto the surface and immersed in a solvent, the transition related to the collapse of the layer caused by a decrease in the solvent strength loses the phase character for uncharged chains, but retains the character of the first-order phase transition for polyelectrolyte chains.

Key words: Monomolecular polymer layers; polymer liquid crystals; orientational ordering; phase transitions; grafted polymer layers; low-dimensional systems

Introduction

According to the character of spatial organization, thin monomolecular polymer layers may be divided into two types: the layers formed by polymer chains lying on the surface (or in a thin layer near the surface) and the layers formed by polymer chains grafted onto the surface at one end and oriented mostly normally to the surface (Fig. 1). From a theoretical point of view the common feature of these systems is their lower effective dimensionality than that of ordinary three-dimensional systems. In the former case the chains in the layer (and the layer as a whole) are two-dimensional, whereas in the latter case the chains in the layer are effectively one-dimensional (the exact meaning of this statement will be discussed below).

It is known that the decrease in the dimensionality of the system always results in the appearance of some anomalies in its behavior. At first, this decrease leads to the changes in the character of phase and non-phase cooperative transitions occurring in these systems.

In the present paper, we consider some typical features of the layers of both types and analyze the phase and non-phase conformational transitions in these layers. The first part of the paper is devoted to the transition of the orientationally ordered state in a planar layer, containing anisotropic (or anisotropically interacting) molecules. In the second part of the paper, we investigate the monolayers formed by polymer chains grafted onto a planar surface at one end and immersed in a solvent. Particular attention will be paid to the analysis of the transition related to the collapse of the layer caused by a decrease in the solvent strength.

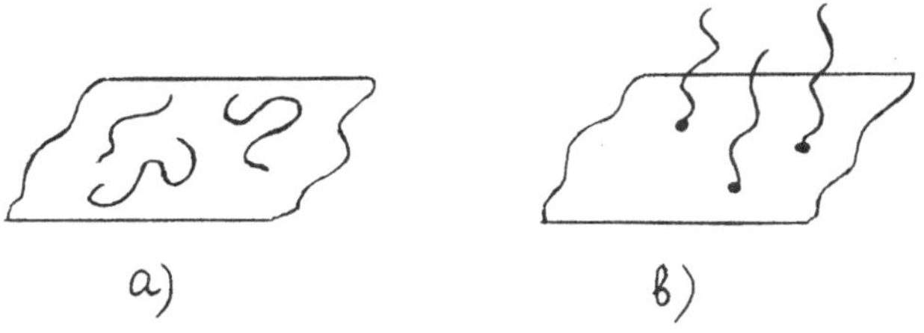

Fig. 1. Two types of layers: a) chains are lying on the surface; b) chains are grafted onto the surface at one end

I. Planar polymer layer; chains lying on the surface

A planar layer of the chains lying on the surface or near it in a thin layer with a fixed thickness (which does not depend on the degree of polymerization of the chains N) can be considered as two-dimensional (2-D) system. A number of quantitative characteristics of such systems differ markedly from those of common three-dimensional (3-D) systems. For instance, in a dilute 3-D solution the asymptotic power dependence of the average dimensions of individual macromolecules R on the degree of polymerization is given by

$$R \sim N^{\nu} . \tag{1}$$

The characteristic indexes are: $\nu_+ = 3/5$, $\nu_0 = 1/2$, $\nu_- = 1/3$ (we use sign "+" for swelling in a good solvent, sign "0" for Θ-solvent, and sign "—" for the collapsed (globular) state below the Θ-point under the condition that aggregation is prevented). For a 2-D dilute solution, numerical values of all the indexes are different: $\nu_+ = 3/4$, $\nu_0 = 4/7$, $\nu_- = 1/2$.

This difference in the value of indices ν_+ and ν_- for the 3-D and 2-D systems is due to the dependence of the density of molecular coil at fixed values of N and R on the dimensionality of the system d, i.e.,

$$\rho \sim NR^{-d} . \tag{2}$$

Therefore, these indexes can be described by the following universal equations

$$\nu_+ = 3/(d + 2) , \quad \nu_- = 1/d . \tag{3}$$

The index ν_0 under the Θ-conditions for the 2-D system also differs from the Gaussian value of ν_0 = 1/2 for the 3-D system, owing to an increase in density with decreasing dimensionality d.

Hence, the conclusion that ν_0 is not equal to 1/2 for the 2-D system is evident enough. However, the determination of the correct numerical value of ν_0 has been a subject of numerous investigations resulting in the wide spectrum of values from ν_0 = 0.505 to ν_0 = 3/4. The value of ν_0 = 4/7 mentioned above has been recently obtained by different approaches by a number of authors (see detailed discussion in [1]).

Now let us consider the main subject of this part of the paper, i.e., the 2-D solution or melt capable of liquid-crystalline orientational (nematic) ordering. Analogous 3-D systems were investigated both theoretically and experimentally for a few decades. We will briefly summarize some useful results of these investigations.

The orientationally ordered state (nematic l.c.) can be generated in systems of anisotropically interacting particles. Two types of nematics may be distinguished with respect to the character of intramolecular interactions: lyotropic l.c., i.e., solutions of molecules of an extended shape in which the nematic phase is formed as a result of steric impermeability of molecules, and thermotropic l.c. for which the main factor of the formation of the ordered phase is the anisotropic van-der-Waals attraction between the molecules. In lyotropic systems, phase transition occurs with increasing concentration in solution, whereas in thermotropic systems it occurs with decreasing temperature of the melt. There are, naturally, two branches of theoretical studies that have sprung from the theories of Onsager [2] and Flory [3] (lyotropic l.c.), and from Mayer-Saupe's theory [4] (thermotropic l.c.).

The results of all these mean-filed investigations of the nematic ordering of 3-D systems may be briefly generalized as follows: the transition from the isotropic into the nematic phase is the first-order phase transition with an abrupt change in the order parameter.

$$S = (3\langle\cos^2\theta\rangle - 1)/2 , \tag{4}$$

(θ is the angle between the director and the chosen axis of the particle) at the transition point from S = 0 in the isotropic phase to S = 0.43 [4] and S = 0.8 [2, 3] at the boundary of the anisotropic phase in thermotropic and lyotropic systems, respectively.

Thermotropic systems

In this paper, we deal with analogous 2-D systems. First, let us determine the order parameter for both 2-D and 3-D systems as follows

$$S = (d\langle\cos^2\theta\rangle - 1)/(d - 1) , \tag{5}$$

where d is the dimensionality of the system. For the 3-D systems definitions (4) and (5) evidently coincide. For a system of anisotropically interacting particles, we use a standard scheme of the mean-field Maier-Saupe's theory [4].

The free energy of a particle $F(p, T)$ in an anisotropic system consists of isotropic $F_{is}(p, T)$ and anisotropic $F_{anis}(p, T, S)$ terms; the latter term is determined by the anisotropy of an angular distribution of particles, i.e., by the form of the function of distribution of angles between the axes of particles $f(\theta)$ and the director, and also by the value of the order parameter S. Hence,

$$F_{anis} = F_{entr}(f(\theta)) + F_{energ}(p, T, S) . \tag{6}$$

The function F_{entr} in Eq. (6) determines the decrease in entropy as a result of anisotropic angular distribution, and

$$F_{energ} = -US^2/2 \tag{7}$$

describes the energetic gain of a particle in the molecular field with a given order parameter S. The condition of self-consistence allows us to obtain the expansion of F_{anis} into a power series of the order parameter S in the vicinity of the point where $S = 0$.

We used this treatment for three different 2-D models [5, 6]: for a system of particles with a continuous distribution of orientations (continuous model), and for two systems with discrete orientations of particles (square and hexagonal latices). For the continuous model, we have the following expansion of F_{anis} into a power series of S [5]:

$$F_{anis}/kT = (1 - U/2kT)S^2 + S^4/4 + \dots . \tag{8}$$

The absence of the cubic term essentially distinguishes expansion (8) from the analogous one for the continuous 3-D model:

$$F_{anis}/kT = 5/2(1 - U/5kT)S^2 - \frac{25}{21}S^3 + \frac{425}{196}S^4 + \dots . \tag{9}$$

In accordance with the general mean-field theory of phase transitions developed by Landau [7], different forms of free energy expansion correspond to different kinds of phase transition. In contrast to the 3-D case, expansion (8) completely satisfies the criterions of the second-order phase transition (the absence of the cubic term in (6) is the most important feature); phase transition occurs at the temperature T_c which is determined by the conversion of the factor at S^2 to zero, hence

$$U/kT_c = 2 . \tag{10}$$

Generally speaking, this result is a consequence of the mean-field theory; actually, the increase in fluctuations distorts the long-range order in a 2-D continuous system, making the existence of the ordered phase impossible [7]. However, because of the weak logarithmic divergence of fluctuations, an order is retained within limited regions, the sizes of which may be relatively large; to a certain extent, this may be an excuse to use the mean-field theories in the studies of 2-D continuous systems.

The 2-D lattice models, i.e., the 2-D systems with discrete symmetry, are of the special interest, because the long-range order is possible for them (it should be emphasized that the discrete character of orientations, but not of the positions of particles is important). The investigation of the expansions of F_{anis} into a power series of S shows that the phase characteristics of the system depend on the kind of latice. For square lattice (the possible orientations of the particles are $\theta = 0, \pm\pi/2, \pi$) the absence of the cubic term in expansion, just as for the 2-D continuous model, leads to the second-order phase transition. In contrast, for the hexagonal lattice ($\theta = 0, \pm\pi/3, \pm 2\pi/3, \pi$) non-zero factor at S^3, as in the case of the 3-D continuous model, Eq. (9), causes the first-order phase transition. The points of phase transition for lattice models are determined by the following relations:

$$U/kT_0 = \begin{cases} 1 \text{ for square lattice} \\ 1.85 \text{ for hexagonal lattice} . \end{cases} \tag{11}$$

Lyotropic systems

As already noted, the nematic phase can be formed because of anisotropic forces of attraction in the thermotropic systems, or as a result of an anisotropic repulsion of molecules of an extended shape in the lyotropic systems. Considering the latter situation, we will restrict ourselves to DiMarzio's simplest lattice model [8] of a solution with discrete orientations of the particles along the lattice axes (q is the number of axes). In an athermal system the free energy is completely determined by the combinatorial term, i.e.,

$$-F/kT = \ln Z \,, \tag{12}$$

where Z is the number of different locations of N_i particles on the lattice in the direction i ($i = 1, 2, \ldots, z$). In the mean-field approximation, the probability of a vacant point is determined by the mean fraction of vacancies; hence, Z is given by

$$Z = \frac{\prod_{i=1}^{z} [M - (p-1)N_i]!}{N_0! \left(\prod_{i=1}^{z} N_i! \right) (M!)^{z-1}} \,, \tag{13}$$

where N_0 is the number of vacancies, M is the number of lattice cells, and $p = 1/D$ is the axial ratio, i.e., the ratio of particle length 1 to its thickness D. Let $i = z$ be the preferential direction and the parameter s be the fraction of particles oriented in all other directions. We define the order parameter S in the following way:

$$S = 1 - zs \,. \tag{14}$$

It can be easily shown that when $z = 2$ (2-D, square lattice) and $z = 3$ (2-D, hexagonal lattice or 3-D, cubic lattice) this definition of the order parameter S coincides with that used for thermotropic systems, Eq. (5). The free energy expansion into a power series of S, Eqs. (12—14), at $z = 2$ is the following:

$$F_{\text{anis}}/MkT = A(S/2)^2 + C(S/2)^4 + \ldots \,, \tag{15}$$

where

$$A = -\frac{\rho'}{1 - \rho'/2} + \frac{2\rho'}{p - 1} \tag{16}$$

$$C = -\frac{\rho'^4}{6(1 - \rho'/2)} + \frac{4\rho'}{3(p - 1)} \,, \tag{17}$$

where $\rho = pN/M$, $\rho' = \rho(p - 1)/p$.

Consequently, the nematic ordering on square lattice proceeds as a second-order phase transition for lyotropic systems, because of the absence of the cubic term in expansion (15). The critical point ρ_c is determined by the condition $A = 0$; hence,

$$\rho_c = 2/(p - 1) \,. \tag{18}$$

In accordance with Eq. (18), the nematic ordering on the square lattice is possible only for particles with an axial ratio $p \geqslant 3$; at $p \leqslant 3$ we have $\rho_c \geqslant 1$, i.e., the point of phase transition has no physical meaning. It should be noted that the exact solution of the problem of the dimers on a square lattice also does not lead to the phase transition into the ordered state [9].

Now the case of a hexagonal lattice will be considered. In this case, the number of orientations on the lattice is $z = 3$, and the problem of the location of particles along the axes of this lattice is completely equivalent to that of the locations of particles along the axes of a cubic lattice in the 3-D systems. The expansion of the free energy at $q = 3$ is the same for 2-D hexagonal and 3-D cubic lattices:

$$F_{\text{anis}} = A(S/3)^2 + B(S/3)^3 + C(S/3)^4 + \ldots \,, \tag{19}$$

where

$$A = -\frac{5\rho'^2}{3 - \rho'} + \frac{9\rho'}{p - 1} \tag{20}$$

$$B = \frac{7\rho'^3}{3(3 - \rho')^2} - \frac{9\rho'}{p - 1} \tag{21}$$

$$C = \frac{9}{2} \left[-\frac{\rho'^4}{(3 - \rho')^3} + \frac{9\rho'}{p - 1} \right] . \tag{22}$$

Hence, according to the mean-field approximation the ordering of a 2-D solution of rod-like particles on a hexagonal lattice, just as that of 3-D solution on a cubic lattice, occurs as the first-order phase transition.

Exact theories

Consequently, the type of phase transition, irrespective of the kind of intramolecular interactions in the system, is determined both by the symmetry of the system and by its dimensionality. On 2-D square lattice the second-order phase transition from the isotropic to the nematic state takes place for both thermotropic and lyotropic systems, whereas on the 2-D hexagonal lattice (just as on the 3-D cubic lattice) we obtained the first-order phase transition. It should be noted that the system with discrete orientations of the elements on the axes of a square lattice is equivalent to the 2-D one-component (double-position) Ising model, the theory of which has been developed by Onsager [10]. In the well-known Ising model spins are oriented normally to the lattice plane, the parallel orientations of the neighboring spins being energetically more favorable than the antiparallel ones; this is the dipole ordered sytem. The liquid crystalline system is the quadruple ordered system. In the case of a square lattice, the system (just as in the Ising model) is double positional and may be considered as a single-component one. The orientational interaction implies that the parallel or antiparallel orientations of neighboring elements are energetically more preferable than the perpendicular orientations. The ordering of the one-component Ising model occurs as the second-order phase transition [11].

On the other hand, it should be mentioned that in the case of the hexagonal lattice, in contrast to the square one, the system is a three positional one. In the terms of the Ising model, it is a bicomponent; but not a single-component model. In the theory of ferromagnetism the Potts lattice model is analogous to this [12]. In the Potts model there are z possibilities for the position of each spin on the 2-D lattice; the energy of intramolecular interactions between the neighboring particles is given by Kroneker's symbol, i.e., $-J\delta(\sigma_i, \sigma_j)$, where

$$\delta(\sigma,\sigma') = \begin{cases} 1, & \text{at } \sigma = \sigma' \\ 0, & \text{at } \sigma \neq \sigma' . \end{cases} \tag{23}$$

At $z = 2$ the Potts model is apparently equivalent to the double-positional Ising model. Exact solutions of the Potts model have been obtained in the vicinity of the critical point on different 2-D lattices; for all of them at $z \leqslant 4$ the phase transition occurs continuously (second-order phase transition), whereas at $z > 4$ it occurs with latent heat (first-order phase transition) [12].

Hence, our conclusion about the dependence of the phase transition on the number of possible orientations on the lattice is in accordance with the results of rigorous theories. However, the mean field approximation used in our research does not predict exactly the critical value of z for which the type of phase transition changes. The theory developed above gives the second-order transition for square lattice only ($z = 2$), while Potts models have the second-order transition at $z \leqslant 4$. This difference is not astonishing, because the mean-field theories generally sharpen phase transition in comparison with the real situation; in particular, mean-field theory gives transition to the ordered phase for one-dimensional system, whereas exact theories show that a completely ordered state is impossible for this system [13].

II. Collapse of grafted polymer monolayers

In this section, we will consider polymer monolayers formed by macromolecules grafted onto an impermeable planar surface at one end and immersed in a solvent (Fig. 1b). Monolayers of this kind are used for the modification of the surfaces of colloid particles in order to prevent flocculation [14]. The sterical stabilization of the dispersion can be essentially improved by the use of charged (polyelectrolyte) chains [15]. Adsorbents with surfaces modified by grafted polymer are used in reverse phase chromatography, etc. In all these applications the dependence of chain conformation and the structure of the modifying polymer layer on the external conditions (solvent strength, temperature, concentration of the added salt, etc.) is of great interest.

The aim of this part of the paper is the consideration of chain conformations in the grafted polymer (including polyelectrolyte) layers over a wide range of solvent strength and ionic strength of the solution. The analysis of the character of the conformational transition related to the collapse of the monolayer caused by the decrease of the solvent strength will also be carried out.

We will consider a layer formed by long (the number of units $N \gg 1$) chains grafted sufficiently densely to ensure the overlapping of neighboring coils: $\sigma \gg R^2$, where σ is the surface area per one grafted chain, and R is the radius of gyrations of a free chain in a solution.

As shown in previous works [16—18] under the conditions of strong overlapping, interchain nonelectrostatic interactions lead to the chain stretching normally to the grafting surface, so that the chain dimensions in this direction scales with N as $H \cong N$ i.e., as in the space with the effective dimensionality $d = 1$ (see Eq. (3)). (Note, that in the lateral direction the chains retain their Gaussian dimensions.) This effective unidimensionality of chains in the grafted layer is manifested in a character of the transition related to the collapse of the layer caused by the decrease of the solvent strength. Indeed, as will be shown below, in contrast to the collapse of an individual chain in a solution (coil-globule transition) which occurs as the second-order phase transition (or the first-order phase transition for stiff chains), the collapse of a layer of chains grafted onto an infinite planar surface is not of the phase character, just as is any cooperative transition in a one-dimensional system (see also [19, 20]).

On the other hand, it will be shown that the charging of the chains in the grafted layer leads to the appearance of effective long-range interaction in the direction normal to the grafting surface. This effective interaction has an entropic origin and is not connected with polymeric chains, but rather with the entropy of mobile counterions. As a result, the collapse of the layer of grafted polyelectrolyte chains acquires the character of jumpwise first-order phase transition, as in 3-D systems.

First, we will consider the conformational transition related to the collapse of the grafted polymer layer caused by the decrease in the solvent strength, i.e., by an increase in the short-range interunit attraction.

The equilibrium structure of the layer is determined from the condition of the minimum of its free energy, ΔF. In our consideration, we will restrict ourselves to the approximation in which ΔF can be presented as a function of the thickness of the layer H alone or of corresponding mean relative characteristics (swelling coefficient α with respect to Gaussian dimensions of the chain, $\alpha = H/N^{1/2}$, or mean volume fraction of units in a layer $\varphi = Na^3/(\sigma H)$) only. (More detailed analysis of the intrinsic layer structure and the structural rearrangement accompanying the collapse of the layer of uncharged chains is contained in [20–22]).

In the approximation used, the free energy ΔF includes several contributions of different physical origins:

$$\Delta F = \Delta F_{\text{conf}} + \Delta F_{\text{conc}} + \Delta F_{\text{trans}} \,. \tag{24}$$

The first term, ΔF_{conf}, describing the losses of conformational entropy caused by the stretching or compression of the chains in a layer with respect to their Gaussian dimensions can be presented [23] in the form

$$\Delta F_{\text{conf}} \cong \begin{cases} \alpha^2 - \ln \alpha^2 \,, & \alpha \geqslant 1 \\ \alpha^{-2} + \ln \alpha^2 \,, & \alpha \leqslant 1 \end{cases} . \tag{25}$$

(Here and below the contributions to ΔF are calculated per one chain and presented in kT units and all numerical coefficients are omitted.)

The term ΔF_{conc} describes the short-range volume interactions in a layer and may be presented as a virial expansion in powers of φ:

$$\Delta F_{\text{conc}} \cong Nv\varphi + Nw\varphi^2 \,, \tag{26}$$

where v and w are the second and the third virial coefficients of interunit interactions, respectively. The change in the solvent strength which can be described as a relative deviation from the Θ-temperature affects the value of the second virial coefficient $v \cong v_0(T - \Theta)/T = v_0\tau$, whereas the third virial coefficient w is virtually independent of T.

The term ΔF_{trans}, which does not contribute to the free energy in the case of an uncharged grafted polymer layer, describes the translational entropy of mobile counterions. These counterions compensate the charge of the chains. An infinite charged plane is known to retain its counterions in a double electric layer of thickness $\kappa^{-1} \cong \varepsilon\tau/(\Sigma e)$ [24], where Σ is the surface charge density, ε is the dielectric constant of the solvent, and e is the elementary charge. If $\kappa^{-1} \ll H$ (we will assume this condition to be fulfilled), almost all counterions are located in the layer, i.e., the layer is electrically neutral as a whole, and we have

$$\Delta F_{\text{trans}} \cong Q \ln(\varphi Q/N) = Q \ln(q\varphi) \,, \tag{27}$$

where Q is the number of counterions per chain and $q \cong Q/N$ is the fraction of charged units. (Note that the value of $q = 0$ corresponds to the layer of uncharged grafted chains).

It can be shown that under the conditions of weakly charged grafted chains ($e^2q^{1/2}/(\varepsilon Ta) \ll 1$) the contribution of Coulombic interaction into ΔF may be neglected.

The minimization of the free energy ΔF (Eqs. (24)–(27)) with respect to α gives the equation for the layer swelling coefficient α as a function of τ:

$$\begin{aligned} a^3 - (Q+1)a &= B\tau + C/a\,, \quad a \geqslant 1 \\ -a^{-1} - (Q-1)a &= B\tau + C/a\,, \quad a \leqslant 1 \end{aligned} \qquad (28)$$

where

$$\begin{aligned} B &\cong v_0 N^{3/2} a^2/\sigma \\ C &\cong w N^2 (a^2/\sigma)^2 \end{aligned} \qquad (29)$$

are the renormalized parameters of binary and ternary non-electrostatic interactions in the layer, respectively.

The analysis of Eq. (28) at $Q = 0$ makes it possible to obtain the power asymptotic dependence for the thickness of the uncharged grafted layer

$$H \cong \begin{cases} Na(v_0 \tau_a^2/\sigma)^{1/3}\,, & \tau \gg |\,\tau^*\,| \\ Na(w^{1/2}a^2/\sigma)^{1/2}\,, & |\,\tau\,| \ll |\,\tau^*\,| \\ Na(wa^2/(v_0\,|\,\tau\,|\,\sigma))\,, & \tau \ll \tau^* \end{cases} \qquad (30)$$

in the region of strong-, Θ-, and poor solvent, respectively. The width $|\,\tau^*\,|$ of the Θ-range (the range of the crossover from the swollen to the collapsed state) is independent of N:

$$|\,\tau^*\,| \cong w^{3/4} v_0^{-1} (a^2/\sigma)^{1/2}\,. \qquad (31)$$

The decrease in τ leads to a smooth (at any N) decrease in the thickness of the layer of uncharged grafted chains. This means that the transition of the layer into the collapsed (globular) state is not of the character of phase transition (see [20] for details).

This character of the collapse of the layer is mainly retained at $Q^2 \ll C$, i.e., when the osmotic pressure of counterions is weaker compared with strong (under the conditions of dense grafting) interchain volume interactions. Under these conditions the layer thickness is still given by Eqs. (30), and the collapse of the layer occurs as a cooperative non-phase transition.

At higher counterion concentration in a layer, $Q^2 \gg C$, the osmotic pressure of counterions predominates over repulsive volume interactions under the conditions of strong- and Θ-solvents and leads to a very strong (polyelectrolyte) swelling of the layer. In these regimes the layer thickness is given by the following equation:

$$H \cong Naq^{1/2}\,. \qquad (32)$$

(Pincus [15] has obtained an analogous result by neglecting the volume interactions in the layer.)

The transition of such a polyelectrolyte layer into the collapsed state occurs with a jumpwise decrease in the layer thickness at $\tau = \tau^{el}$, where

$$\tau^{el} \cong -q^{1/2} w^{1/2}\,; \qquad (33)$$

(Note that $|\,\tau^{el}\,| \gg |\,\tau^*\,|$ at $Q^2 \gg C$).

Hence, the counterion osmotic pressure (counterion entropy) causes not only the strong swelling of the layer, but also leads to the appearance of additional effective long-range interactions of non-polymeric nature in a quasi-one-dimensional polymeric system. At $Q^2 \gg C$ ($q \gg w^{1/2}a^2/\sigma$) the contribution of counterion entropy is sufficient for changing the character of conformational transition, which is related to the collapse of the layer (caused by a decrease in the solvent strength) and, hence, it becomes the first-order phase transition. Note that the effect of jumpwise collapse has previously been known for "dilute" polyelectrolyte networks [25].

The collapse of a swollen polyelectrolyte layer may be caused not only by a decrease in the solvent strength, but also by an increase in the ionic strength of solution. If the concentration of salt in the bulk of the solution c_s is fixed, the equilibrium layer thickness and salt concentration φ_s in the layer are determined by the following conditions:

$$-(\partial \Delta F/\partial a)_{\varphi_s} = \pi^{bulk} \sigma N^{1/2} a \qquad (34)$$

$$(\partial \Delta F/\partial(\sigma H \varphi_s))_a = \mu_s^{bulk}\,, \qquad (35)$$

where π^{bulk} and μ_s^{bulk} are the osmotic pressure and the chemical potential of salt in the bulk of the solution.

The analysis of Eqs. (34) and (35) shows that, at low salt concentration ($c_s \ll q^{1/2}/\sigma$), the layer dimensions weakly depend on this concentration. An increase in salt concentration up to the values of $c_s \gg q^{1/2}/\sigma$ results in a smooth decrease in the layer thickness according to the equation

$$H \cong Nq^{2/3} c_s^{-1/3} \sigma^{-1/3}\,, \qquad (36)$$

up to the value of H determined by Eq. (30). Hence, there is no phase transition. A similar result was obtained on the basis of a slightly different approach by Pincus in [15].

References

1. Birshtein TM, Buldyrev SV (1991) Polymer: in press
2. Onsager L (1949) Ann NY Acad Sci 51:627—639
3. Flory PJ (1956) Proc Roy Soc (London) A234:73—89
4. Maier W, Saupe A (1959) Z Naturforsch 14a:882—889
5. Merkurieva AA, Medvedev GA, Birshtein TM, Gotlib YY (1990) Vysokomol Soedin A32:961—963; (1990) Polymer Science USSR 32
6. Merkurieva AA, Birshtein TM (1991) Vysokomol Soedin A33:141—147; (1991) Polym Sci USSR, 33
7. Landau LD, Lifshitz JM (1976) Statistical Physics Part 1, Nauka, Moscow
8. DiMarizo E (1961) J Chem Phys 35:658—663
9. Kastelyn PW (1961) Physica 27:1209—1212
10. Onsager L (1944) Phys rev 65:117—121
11. Potts RB (1952) Proc Camb Phil Soc 48:106—109
12. Baxter R (1982) Exectly Solved Models in Statistical Mechanics, Academic Press, NY
13. Kac M, Ulenbeck GE, Hemmer PC (1963) Math Phys 4:216—221; (1964) Math Phys 5:60—65
14. Napper DH (1983) Polymer Stabilization of Colloidal Dispersions, Academic Press, London
15. Pincus P (1991) Macromolecules 24:2912—2919
16. Alexander S (1977) J phys 38:983—988
17. de Gennes PG (1980) Macromolecules 13:1069—1075
18. Birshtein TM, Zhulina EB (1983) Vysokomol Soedin A25:1862—1868; (1983) Polymer Science USSR 25; (1984) Polymer 25:1435—1461
19. Borisov OV, Zhulina EB, Birshtein TM (1988) Vysokomol Soedin A30:767—773
20. Zhulina EB, Borisov OV, Pryamitsyn VA, Birshtein TM (1990) Macromolecules, in press
21. Skvortsov AM, Gorbunov AA, Pavlushkov IV, Zhulina EB, Borisov OV, Pryamitsyn VA (1988) Vysokomol Soedin A30:1615—1622; Polymer Science USSR 30:1706—1715
22. Zhulina EB, Pryamitsin VA, Borisov OV (1989) Vysokomol Soedin A31:185—193; Polymer Science USSR 31:205—216
23. Birshtein TM, Pryamitsyn VA (1987) Vysokomol Soedin A29:1858—1864; (1991) Macromolecules 24:1554—1560
24. Israelachvili JN (1985) Intermolecular and Surface Forces, Academic Press, London
25. Vasilevskaya VV, Khokhlov AR (1982) in Mathematical Methods for Polymer Investigations, Ed. SCBI, Poustchino, USSR

Received January 15, 1991
accepted February 8, 1991

Authors' address:

Prof. T. M. Birshtein
Institute of Macromolecular Compounds
Academy of Sciences of the USSR
199004 Leningrad, USSR

Discussion

WENDORFF:
How does the transition occur, proceeding for the grafted chain, from a concentrated tc a dilute state: continuously or stepwise?

BIRSHTEIN:
Continuoulsy.

KILIAN:
And, would you comment about the role of chain-end-defect-energies?

BIRSHTEIN:
In this theory, we suppose that there is no extra energy of interaction connected with end groups. If the end groups differ from the other groups inside the chains, the chain end distribution becomes a function of this extra energy.

KILIAN:
Would you expect asymptotics when increasing the number of chains of different length?
BIRSHTEIN:
We obtain such asymptotics of layer-density distribution; the form of asymptotics depends on the distribution of length of grafted chains.

KILIAN:
Could you comment about the role of the coordination number?

BIRSHTEIN:
In the mean field approximation, the order of the phase transition from isotropic to nematic state changes with going from quadratic to hexagonal lattice, because of the change of the symmetry of the system.

Progress in Langmuir-Blodgett-electronics

L. Brehmer

Institute of Polymer Chemistry "Erich Correns", Teltow-Seehof, FRG

A general trend in the micro-electronics is the miniaturization of (lateral) structures toward the natural limit of the material, that means to molecular level.

The LB-technique is a practicable input in the molecular electronics. Therefore, we focus the attention in the lecture on the progress of the electrical and electronic effects and their phenomenological description.

Although the electric effects of LB-films are studied thoroughly, until now there is no large scale production or applications of LB-films. The reasons for this are:

i) the preparation of high-quality LB-films,
ii) the understanding of the electrical effects,
iii) experimental problems (electrodes).

The present state of art of the following topics will be discussed in the lecture:

i) specific aspects of LB-films,
ii) large scale production of LB-films,
iii) special electric effects,
iv) potential applications,
v) conclusions.

Author's address:

L. Brehmer
Institute of Polymer Chemistry "Erich Correns"
O-1530 Teltow-Seehof, FRG

Progress in Colloid & Polymer Science Progr Colloid Polym Sci 85:47—51 (1991)

Deposition and investigation of protein Langmuir-Blodgett films

V. Erokhin and L. A. Feigin

Institute of Crystallography Academy of Sciences of the USSR, Moscow, USSR

Abstract: The protein Langmuir-Blodgett films can be divided into five groups by different deposition techniques. These films of membrane fragments, membrane proteins, immobilized proteins, water soluble proteins, and modified parts of proteins are discussed along with their properties by the interpretation of special examples.

Key words: Langmuir-Blodgett films; proteins; protein films; membranes

The Langmuir-Blodgett (LB) technique is a powerful tool to investigate models of biological membranes and for studying lipid-protein interactions [1, 2]. The method also allows the creation of complicated molecular systems, i.e., to realize ideas of molecular architecture [3]. As specific feature of the process is that it proceeds at room temperature and atmospheric pressure, which are convenient conditions for proteins.

Langmuir and Shaefer were the first to work with protein LB films [4].

A new wave of interest in such arose a few years ago from the possibility of using protein LB films as sensing elements in biosensors [5, 6].

It is possible to divide all the protein LB films into five groups based on differences in deposition technique, namely:

— films of membrane fragments;
— films of membrane proteins;
— films of immobilized proteins;
— films of water soluble proteins, and
— films of modified parts of proteins.

Let us consider each group of proteins in detail.

LB films of membrane fragments

LB films of bacteriorhodopsin are a typical example of this group. Until recently this protein has been one of the most popular subjects of the LB technique [7—9]. In this case, one uses a water solution of purple bacteria fragments. Then the solution is spread on the surface of a water subphase. It should be mentioned that in such a process a lot of protein molecules penetrate the water subphase. This fact indicates that we initially have a solution of three-dimensional aggregates, mostly hydrophilic, tending to be dissolved. Such a leakage can be diminished by using a dense salt solution as a subphase [10].

But the main disadvantage of this process is that salt particles are desposited together with the bacteriorhodopsin in the film. The presence of the salt particles results in a very low ordering in such films. It is possible to increase the ordering by carefully washing the sample after each monolayer deposition (this experiment was performed together with G. Sukhorukov and will be published later).

Another approach consists in using a hexane solution of the fragments [11]. Such an approach probably slightly diminishes the leakage of the protein from the surface, but in any case, one compresses not individual molecules, but performed domains. As a result there are specific defects in the LB films: cracks between the domains.

Films of membrane proteins

Membrane proteins are extremely suitable for use in LB technology. Such proteins contain a large hydrophobic part in addition to hydrophilic areas. The presence of a hydrophobic part keeps the protein at the water surface and allows to reach a high surface pressure (up to 50—60 mN/m). The area per molecule is close to the cross-section of the protein molecule calculated from crystallographic data in this case [12].

It should be mentioned that during isolation from membranes protein molecules are treated by detergents. Hydrophobic areas of the proteins become screened by detergent molecules after such a procedure. The detergent-protein complex contains up to 30% of detergent [13] and is water soluble. Detergent molecules are suitable to be transfered from the hydrophobic area of the protein to the subphase surface after speading at a water surface. So protein molecules form a monolayer and detergent molecules fill the space between them.

LB films of reaction centers (RC) of purple bacteria are a typical example of this group. In membranes the protein carries out charge separation following absorption of light.

There is a good review by Tiede [1] presenting data about optical and electrical properties of RC LB films. Here, we mostly consider the structural parameters of such films.

LB films of RC from *Rhodobacter sphaeroides* were studied by comparing structural parameters of the film at each step of its formation with those of the protein molecule itself [12]. For this reason small-angle x-ray scattering of the protein was done in a solution before making the LB films. The study has allowed us to determine the shape and size of the protein globules. The molecules were approximated as a cylinder with a diameter of d = 35 Å and a height of h = 75 Å. The middle of the cylinder is surrounded by a belt of detergent molecules. The external diameter of the complex is d = 50 Å.

The π-A isotherm of the protein monolayer allows the calculation of the area per molecule. In the case of RC the value was found to be equal to 2500 $Å^2$. The monolayer thickness turns out to be D = 86 Å (from x-ray data). A comparison of the film parameters with sizes of the protein globule permits us to suppose that the molecules vertically stand in the film (the long axis and electron transfer chains are perpendicular to the substrate). The slight increase in the monolayer thickness with respect to the largest dimension of the protein molecule is due to the presence of water interlayers between the hydrophilic regions of the joined monolayers. The coincidence of structural parameters indicates the absence of denaturation and this was confirmed by electro-optical measurements with sandwich-structures [14] and MIS transistor [12].

As mentioned above, all the protein molecules in film stay vertically, but an excess of one orientation to the opposite is not big (12%) [14]; this fact is due to closed properties of the hydrophilic sides of the molecule.

To increase the anisotropy of RC LB film, one should use a more anisotropic protein molecule; RC from *Chromatium minutissimum* is such a molecule. It contains an additional subunit — cytochrome C, thus, one side of the molecule is more hydrophilic than the other. The anisotropy is confirmed by the fact that it is impossible to transfer the LB film of RC *Chromatium minutissium* from the water surface to a substrate [15]. This is due to the strong interaction of the cytochrome subunit with water. Cytochrome C contains a positive charge (8^+) at pH 7.0 [16]. Therefore, the film can be deposited only onto a charged substrate. In order to deposit such films we carried out the following procedure. A hydrophobic substrate was moved downward through a monolayer of arachidic acid. At neutral pH the head-groups of the acid are in a deprotonated form and contain a negative charge. During upward motion the substrate goes through the RC monolayer and the negative head-groups of the acid bind to the positive cytochrome side of the proteins. Repeating the process, it is possible to deposit a protein-containing superlattice. The x-ray pattern of the superlattice is presented in Fig. 1. A spacing of the elementary unit (monolayer of arachidic + monolayer of RC) is equal to 115 Å and is in good correlation with the length of both molecules. It should be mentioned that interlayers of arachidic acid make the film more ordered. The x-ray pattern of a monocompound LB film of RC from *Rhodobacter sphaeroides* is presented in Fig. 2 for comparison.

Films of immobilized proteins

The most interesting example of this kind of film is that of the glucosoksidase (GOD) molecule surrounded by 200 molecules of charged lipid [17]. This complex was dissolved in benzol. A stable monolayer was formed at a water surface (π up to 50 mN/m) by the usual technique.

At the water surface the lipid surrounding opens and one has a lipid monolayer on water with bound GOD molecules under the water. Protein molecules

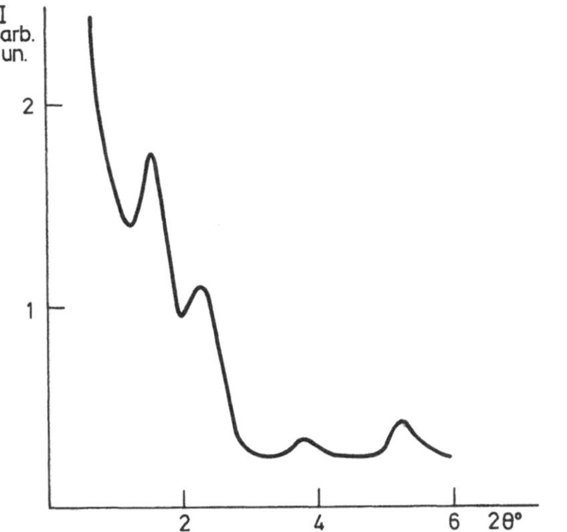

Fig. 1. X-ray pattern of LB superlattice containing 15 periods of arachidic acid — RC from *Chromatium minutissimum* alteration

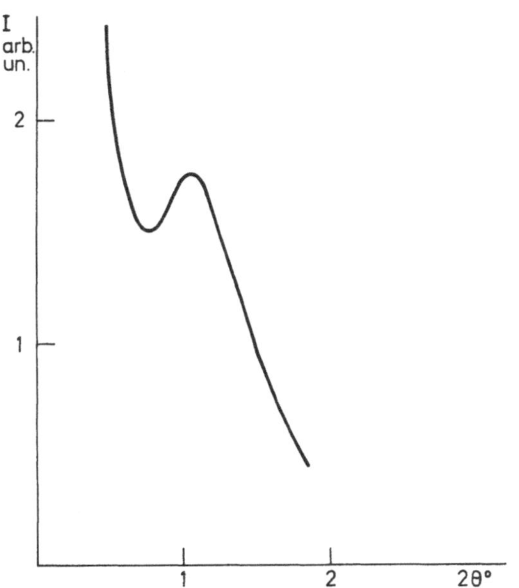

Fig. 2. X-ray pattern of LB containing 30 monolayers of RC from *Rhodobacter sphaeroides*

remaine in the monolayer for at least 6 h without becoming denatured. The monolayer was transferred to a substrate at π = 40 mN/m. A platinum electrode was used as a substrate. The electrode with the LB film of GOD was placed into an electrochemical cell (pH 5.6). An electrical signal appeared after 5 s when glucose was added to the solution. The value of the current was linearly proportional to the glucose concentration.

Films of water soluble proteins

This group seems to be most difficult for the LB technique. There are two reasons for this. The first one is that such proteins are water soluble and so it is not easy to make a monolayer at a water surface. In principle, it is possible to use more dense substances as a subphase. Mercury was used for this reason [18]. But if one uses a dense subphase, the second problem arises: denaturation of the proteins can take place at the subphase surface. Taking into account their behavior in a monolayer, water soluble proteins can be divided into two subgroups. In order to distinguish them it is useful to draw a dependence as presented in Fig. 3. This is the temporal dependence of the area inside barriers when a constant surface pressure is maintained. The dependence for RC monolayers can be considered as a reference line. The presence of large hydrophobic areas keeps RC molecules at the water surface.

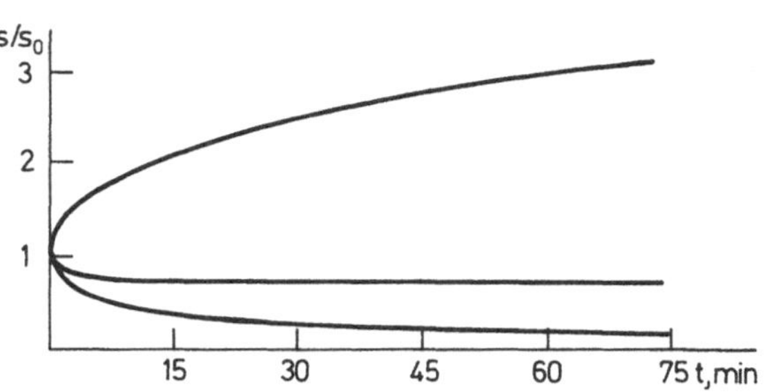

Fig. 3. Temporal dependence of the area inside barriers when a constant surface pressure is maintained (π = 20 mN/m)
1) Cytochrome C monolayer;
2) RC monolayer;
3) Immunoglobuline G monolayer

There are two types of water soluble molecules. The behavior of proteins from the first subgroup is similar to that for the cytochrome C monolayer. The area inside the barriers increase in time. Such dependence indicates the fact that the polypeptide chains of the molecules enlarge, i.e., denaturation takes place. This means that it is impossible to apply the LB technique in its original version to such proteins. But there is another subgroup of water soluble proteins, the behavior of which is similar to those for an immunoglobuline monolayer. The area inside the barriers diminishes in time. This indicates that denaturation does not take place; this is likely due to the presence of covalent *s-s* bridges inside the molecule which are much stronger than the polar interaction of the molecular surface with water molecules. As the molecule is watersoluble, there is some leakage of the protein from the surface into the volume of the subphase. It is possible to apply the LB technique in its original version to such proteins. Let us take immunoglobuline G LB films as an example of this subgroup because of their significant practical importance.

First of all, it should be mentioned that it is impossible to obtain any numerical information from the π-A isotherms of such proteins. This is due to leakage of the protein molecules. But as the area per molecule is important, one should find another measurement permitting its calculation. We propose to use microgravimetric measurements obtained from quartz resonators [19]. Before making measurements of the area per molecule, it is necessary to smooth the surface of the resonators in order to obtain reproducible values. For IgG molecules the area per molecule was found to be equal to 3200 $Å^2$ [19].

It is possible to increase the mechanical stability of the immunoglobuline monolayer by injecting glutar aldehyde into the water subphase under the monolayer [20].

LB films of immunoglobulines can be used as active elements in biosensors. An ability to bind specific antigens is used in such devices. Detection of this binding can be done by several methods. It is possible to use optical [21], electrical [22], and piezoelectrical [6] methods. Experimental curves for a piezoelectric immunosensor are presented in Fig. 4. The dependence illustrates the fact that immunoglobuline molecules do not lose their activity in LB films, and it is possible to distinguish the specific binding from the background level.

But what should one do if it is necessary to make LB films of water soluble proteins from the first subgroup? In this case it is possible to use two techniques based on different phenomena, but with a similar realization. The first technique involves a charge interaction [23, 24]. A monolayer of

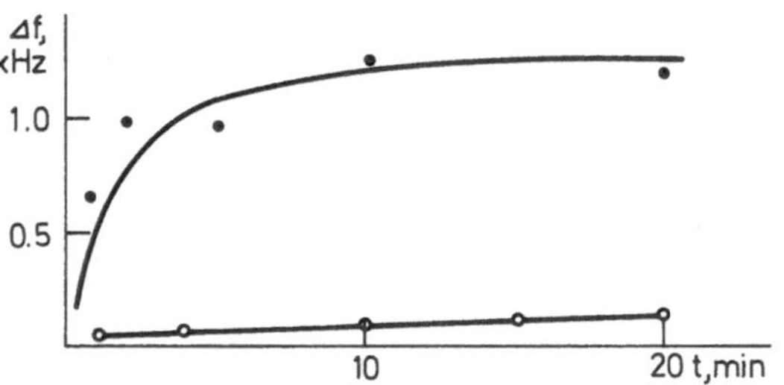

Fig. 4. Experimental curves of immunosensor based on quartz resonators (active element — LB film of antibody, in a solution — protein antigen)
1) Specific interaction;
2) Nonspecific background

charged lipid is formed on the water surface. Then a solution of proteins with a charge opposite to those in the lipid head groups is injected into the subphase. Protein molecules bind electrostatically to the lipid monolayer. It is possible to deposite such a complex monolayer onto a substrate.

The technique was applied to create cytochrome C containing LB films [23]. It was found by polarization spectroscopy that protein molecules are oriented — porphyrin rings are arranged parallel to the substrate plane. The other technique is based on specific binding interactions. Let us consider the following example [25]. A monolayer of the dipalmitidil derivative of biotin, which is a substrate for streptavidin, is formed at the water surface. Fluorescence labeled streptavidin solution is injected into the subphase. Two-dimensional domains of streptavidin are formed at the surface after approximately 1 h. The orientation of protein molecules in one domain turns out to be the same (by data of fluorescence polarization).

There is a tool, named the Fromherz through, which is very useful for realizing both techniques. This is a multisectional circular through with the sections being separated by movable walls [26]. The first section is the site of monolayer formation. This monolayer is then transferred by the barriers of the section containing the protein solution. Charge or specific binding interactions can take place in this section. Then this complex monolayer is subsequently transferred to a section with clean water and to a section with a subphase convenient for deposition to the substrates. Other versions of the Fromherz trough are possible, but it seems to be necessary to separate the subphase volumes for monolayer formation, reaction, and deposition.

Films of modified parts of proteins

It seems to be best to separate from the protein molecule the parts containing active centers and to bind them to long hydrophobic chains. Such modification will diminish protein leakage and will allow the formation of an anisotropic oriented monolayer, suitable for transfer to a substrate.

F_{ab} fragments of IgG were separated [27]. These fragments contain antigen binding centers. After modification the resulting molecule was:

$$F_{ab}\text{—S—S—}(CH_2)_2\text{—CO—NH—}(CH_2)_{15}\text{—}CH_3 \ .$$

Such a modification stops the penetration of the protein molecules into the volume of the subphase, but does not diminish the antigen-binding activity.

As one can see, there are no practical limitations to the creation of protein LB films. The method is useful both for the production of ultrathin oriented protein films on a solid substrate surface and for the investigation of lipid-protein interactions on the water surface.

Acknowledgement

The authors thank Dr. Yu. Lvov for useful discussions.

References

References may be requested from the author.

Authors' address:

V. Erokhin
Institute of Crystallography
Academy of Sciences of the USSR
Leninsky pr. 59
USSR-117333 Moscow

Discussion

KILIAN:
Have you drawn conclusions about the state of order from the analysis of SAXS-pattern?

EROKHIN:
Length of ordering is possible to calculate from the half-width of Bragg reflections. In the case of protein LB films, the value is usually 3—4 monolayers.

KREMER:
Did you control the relative humidity of the combined LB films? Did you find a humidity-effect on the absorption spectra in the case of bacterio rhodopsin?

EROKHIN:
In our experiments, the humidity did not influence either ordering or spacing of the films.

WARTEWIG:
Would you comment on the sensitivity of the piezoelectric resonator applied? How do you get quantitative information about monolayer by using the piezoelectric resonator method?

EROKHIN:
The sensitivity of the frequency measurements was 10 Hz (non-stability of the resonator). Frequency-shift directly corresponds to the deposited mass. Taking into account the value of covered area, it is possible to calculate the area per molecule.

HAVRANEK:
Can you give us more information about the influence of crosslinking on the layer thickness?

EROKHIN:
Crosslinking results in diminishing the thickness of IgG LB film. It was registered by SAXS and ellipsometry measurements.

KILIAN:
Would you not mind that your chains are tilted, due to being packed more densely together?

EROKHIN:
It is necessary for hydrocarbon chains to be closely packed for LB film formation. If the head-group of the molecule is large, the chains should be tilted.

Progress in Colloid & Polymer Science Progr Colloid Polym Sci 85:52—54 (1991)

The organization of aliphatic chains in ultra-thin layers and its importance for layer properties

I. R. Peterson and H. Möhwald

Johannes-Gutenberg-Universität, Institut für Physikalische Chemie, Mainz, FRG

Monolayers of amphiphilic monolayers on a water surface have been studied since the turn of the century for their fundamental scientific interest, for their importance to biology, and for their used in the formation of Langmuir-Blodgett films. For most of the intervening period there have been differences of scientific opinion about their molecular organization. Until very recently these opinions could not be backed up by full experimental evidence, because there were very few techniques known which could distinguish the very small quantities of monolayer material from the much larger quantities of water. This has led to an unusual situation where certain pictures of monolayer organization are widely accepted, in particular those due to Adam [1] and due to Harkins [2], while their origins and mutual contradictions are largely forgotten.

A number of recently developed techniques have completely changed this position. Dye molecules used as molecular-scale probes have provided incontrovertible evidence about the nature of phase transitions in the layers, and about the presence of long-range order in the individual phases [3, 4]. Even more explicit data about order in the phases has been provided by x-ray reflection [5] and diffraction [6]. For specific questions these main techniques have been complemented by a range of others [7]. There are advantages of transferring the monolayer to a solid substrate [8—10].

One of the interesting consequences of these new techniques is the new life they have given to the oldest technique of them all: the surface-pressure-area isotherm. Although the interpretation of isotherm results has always been a problem, the new x-ray results have shown that even quite small kinks and changes of gradient correspond to genuine changes of molecular organization, and have confirmed the almost unknown work of Ställberg-Stenhagen, Stenhagen and Lundquist [11—13]. This older but much faster technique has now been used to demonstrate that the different molecular packings so far determined do not occur haphazardly, but conform to a regular pattern as the chain length is changed [14].

With this new lease of life, isotherms have also been used to investigate the phase relationships between materials with different head groups [15].

It is now possible to adjudicate in the debates between Adam, Harkins and Dervichian: of the three, Dervichian was the most nearly correct with his suggestion that one of the monolayer phases must be liquid crystalline. Five liquid crystalline phases have now been identified and assigned to known smectic categories [6], as shown in Table 1.

Table 1. Structures of the 8 well-established monolayer phases

Monolayer phase (Harkins-Stenhagen nomenclature)	State	Smectic category
G	Gas	—
L_1	Isotropic 2D liquid	S_A
L_2	Hexatic mesophase	S_I
L_2'	Hexatic (2 subph.)	S_F/S_{HH}
L_2''	Hexatic mesophase	S_{KH}
LS	Hexatic mesophase	S_{BH}
S	Hexatic mesophase	S_{EH}
CS	True 2D crystal	S_{EC}

References

1. Adam NK (1941) The Physics and Chemistry of Surfaces, 3rd Edn, Oxford: London
2. Harkins WD, Copeland LE (1942) J Chem Phys 10:272
3. Miller A, Möhwald H (1987) J Chem Phys 86:4258
4. Moy VT, Keller DJ, Gaub HE, Mc Connell HH (1986) J Phys Chem 90:3198
5. Kjaer K, Als-Nielsen J, Helm CA, Tippmann-Krayer P, Möhwald H (1989) J Phys Chem 93:3200
6. Kenn RM, Böhm C, Bibo AM, Peterson IR, Möhwald H, Kjaer K, Als-Nielsen J, J Phys Chem (1991) 95:2092
7. Knobler CM (1990) In: Prigogine I, Rice SA (eds) Advances in Chemical Physics, Wiley, New York, p 397

8. Lösche M, Rabe J, Fischer A, Rucha BU, Knoll W, Möhwald H (1984) Thin Solid Films 117:269
9. Bibo AM, Peterson IR (1989) Thin Solid Films 178:149
10. Peterson IR, Steitz R, Krug H, Voigt-Martin I (1990) J Phys France 51:1003
11. Ställberg-Stenhagen S, Stenhagen E (1945) Nature 156:239
12. Stenhagen E (1955) In: Braude EA, Nachod FC (eds) Determination of Organic Structures by Physical Methods, Academic, New York, p 325
13. Lundquist M (1971) Chem Scripta 1 5:197
14. Bibo AM, Peterson IR (1990) Advanced Materials 2:309
15. Bibo AM, Knobler CM, Peterson IR, J Phys Chem (1991) 95:5591

Authors' address:
I. R. Peterson
Johannes-Gutenberg-Universität
Institut für Physikalische Chemie
Jakob-Welder-Weg 11
6500 Mainz, FRG

Discussion

STAMM:

I wonder whether you see a correspondence of the fluorescence microscopy investigations with the x-ray reflectivity and scattering results, in particular, in the region of coexisting phases. You should get a superposition of, for instance, the diffraction peaks of the pure phases, while the reflectivity data will be difficult to interpret.

PETERSON:

Yes, definitely. For DMPA, DMPE, and DLPE monolayers in the L_1—L_2 phase coexistence region, we have fitted the reflectivity profiles, assuming incoherent superposition of the contribution from each phase. With docosanoic monolayers, we have seen diffraction patterns with superimposed peaks from the coexistence regions at the L_2''-CS, L_2-L_2' and S-LS transitions, as well as at the monolayer-bulk transition (collapse), but we have not yet fully analyzed these, preferring to work with the simpler diffraction patterns first.

KRYSZEWSKI:

Are there dye impurities which change the order in an important way in the systems?

PETERSON:

Yes, this is a possibility which must always be kept in mind in the fluorescence microscopy work. A dye probe has a molecular shape which is inherently different from that of a lipid molecule, and there is always a solubility limit beyond which the order changes. Beyond a second threshold, the system becomes once more homogeneous, but with a different packing symmetry. In the systems we work with, to be confident that we are looking at intrinsic monolayer properties and not some artefact of the probe, we use a range of dyes at several different concentrations. For most of them, we have also confirmed the fluorescence results using independent techniques, for example, electron microscope imaging using the charge decoration technique.

BIERSTEIN:

Why are the positions of the microphases so well ordered (referring to the hexagonal packing of circular domains of L_2 phase in a DMPE monolayer at the L_1-L_2 transition)?

PETERSON:

This phenomenon has been the subject of many papers since it was first reported in 1984, and the explanation must be subdivided into two parts. Firstly, there is no doubt that the tendency for the domains to remain apart is due to the electrostatic repulsion between the dipole moments of the lipid molecules normal to the water surface. In the L_2 phase, the dipole moment per unit area is significantly higher than in the L_1 phase. The second aspect, which is not so well understood, is the uniformity of domain diameter. This may be an equilibrium phenomenon and may also be explained by electrostatic interactions, or it may be due to the kinetics of domain growth.

BIERSTEIN:

What about the conformations of macromolecules?

PETERSON:

We can now state unequivocally that only the *G* (gas) and L_1 (liquid expanded) phases have an appreciable amount of conformational disorder, which probably consists in g^+g^- kinking of the aliphatic chains. In the other phases, almost all the molecules are in the all-trans conformation and locally display dense crystalline packing with a correlation length of translational order along the chain essentially equal to the monolayer thickness.

KILIAN:

In the case of a monolayer with impurities, phase coexistence is no longer neutral, and at each temperature there is a well-defined mass-fraction of each phase. Have you proven this phenomenon?

PETERSON:

Yes, this question has been investigated most thoroughly using the fluorescence microscopy technique, using image processing to determine the areal fraction of each phase, for example, in J. Coll. Interface Sci. 126 (1988) 432. In J. Chem. Phys. 86 (1987) 4258, it is shown that the presence of impurities accounts for all observed aspects of the finite temperature range of phase coexistence.

KREMER:

Did you measure the conductivity of the monolayers using a symmetric arrangement of the electrodes?

PETERSON:

I mentioned these conductivity studies to indicate that the results of our monolayer structure investigation have a wider relevance. They were carried out by the group of Tredgold in Lancaster, using an experimental setup very similar to the one also used by Kuhn, Sugi, and Roberts. The cells were not symmetrical. The base electrode was aluminium covered with a native oxide layer, while the top electrode was gold.

KREMER:

Did you observe a phase dependence of the measured conductivity, as I would expect?

PETERSON:

Firstly, the results about monolayer phases which I have just told you about are brand new, and refer to the fatty acids and esters. All published conductivity results refer to the cadmium soaps, whose phase diagram is as yet unknown. There is no way of knowing whether phase differences can explain the large differences in conductivities reported by different groups. We are planning to carry out such measurements, but the time is not yet ripe. Secondly, while I agree that there should be a dependence on phase behavior, I think it may be more complicated than your question implies. The film conductivity appears to be associated with defects which are not in equilibrium on the water surface, but which survive the phase transition from the L_2 to L_2' phase and back. We showed last year in Thin Solid Films 178 (1989) 81 that the rate of defect recombination was phase-dependent.

Accurate measurement of the interface width and density profile between polymers by reflectivity techniques

M. Stamm

Max-Planck Institut für Polymerforschung, Mainz, FRG

Abstract: The application of x-ray and neutron refelectometry for the determination of the interface width and density profile between polymers is described. As an example, the initial stages of interdiffusion between deuterated and protonated polystyrene and the interfacial region between the incompatible materials polystyrene/poly-para-bromostyrene are discussed. Depending on sample preparation, an accuracy in the determination of the interfacial width as low as 0.2 nm can be achieved, which is hardly met by other techniques. Also, the shape of the density profile can be obtained in favorable cases by model fits to the reflectivity curves.

Key words: Interfaces; interdiffusion; blends; x-ray reflectivity; neutron reflectivity

The interface between polymers enters into many pracical applications of polymeric materials. It is particularly important for blends of incompatible polymers, but also for the welding or healing process of miscible materials. Its detailed knowledge is, however, also needed for a comparison with theoretical predictions of the phase separation of blends. The development of the interfacial region with time between miscible materials does, on the other hand, yield valuable information on the nature of the initial stages of the interdiffusion process in polymers. In both cases it is necessary to use a technique for the determination of the interface width which has a resolution much smaller than the radius of gyration of the molecules, which in many practical cases is of the order of 10 nm. Most common techniques for the determination of the interfacial region thus cannot be used.

Forward recoil spectroscopy (FRS) has a resolution of approx. 80 nm [1], secondary ion mass spectrometry (SIMS) of the order of 13 nm [2], nuclear reaction analysis (He^3-NRA) approx. 15 nm [3] and infrared microscopy (IR) of the order of some micrometers [4]. Other techniques like x-ray photoelectron spectroscopy (XPS/ESCA) or scanning tunneling microscopy (STM) are only surface sensitive and cannot be used for depth profiling of a "hidden" interface within the sample.

Recently, x-ray [5, 6] and neutron reflectometry [7, 8] have also been used for polymers to investigate with high precision polymer thin films. The technique is described in different reviews [9—11], and data analysis is very similar to the case of the reflection of light [12]. For the experiments, one needs a thin polymeric film or layer system on a substrate. Requirements on homogeneity, thickness variation and roughness at interfaces are very stringent, but can be met by polymeric materials. We, thus, usually prepare thin films by spin-coating on a flat glass substrate and thus achieve a very good sample homogeneity ($\simeq 10\%$), thickness variation ($\simeq 2\%$), and surface roughness ($\simeq 1$ nm). A typical sample size for x-ray reflectometry is $7 \cdot 2\ cm^2$, and $10 \cdot 10\ cm^2$ for neutron reflectometry.

Experiments are performed on our x-ray reflectometer at Mainz [5] using a rotating anode source at a wavelength of 0.154 nm. For neutron reflectometry measurements, we use the reflectometer TOREMA at Jülich [13], operating at a neutron guide with fixed wavelength of 0.43 nm. While x-ray experiments can only be performed with different polymers where a sufficient electron density difference is present the neutron technique can be used much more generally since a contrast can always be achieved by the deuteration of one component.

We have investigated the interfacial width in two systems. For the immiscible pair — polystyrene (PS) and poly-para-bromostyrene (PBrS) — a very thin interfacial layer is formed during annealing which reaches an equilibrium thickness after some time and does not change further. Deuterated polystyrene (PS(D)) and normal polystyrene (PS(H)) of molecular weight $M_w \sim 700\,000$ g mol^{-1} are miscible and form an increasing interfacial region during annealing. The time-dependence of the interface width shows, however, a characteristic behavior at small annealing times before the usual $\sqrt{t}$-dependence at later times is reached.

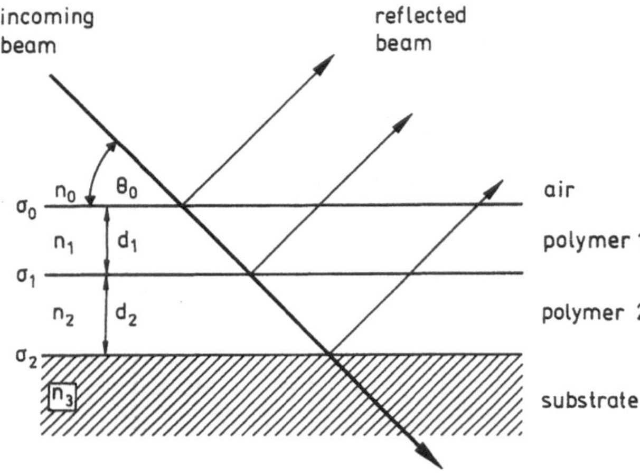

Fig. 1. Schematic diagram of experimental set-up for a reflectivity experiment to determine the width σ_1 of a buried interface. The indices of refraction of air, polymer 1, polymer 2, and the substrate are given by n_0, n_1, n_2, and n_3, respectively. The surface roughness is expressed by the variance of the error-function density profile σ_0, while the interface widths of the polymer 1/polymer 2 and polymer 2/substrate interface are given by σ_1 and σ_2. The beam hits the sample at an incident angle Θ_0 and is partially reflected at the various interfaces. d_1 and d_2 are the film thicknesses of the two polymer films. The effect of refraction is neglected in this figure

A typical experimental set-up is shown in Fig. 1. The sample consists of two thin polymer films on a glass substrate and is hit by the x-ray or neutron beam from the air side at small angles. The reflected intensity is monitored as a function of angle and shows a characteristic modulation due to the interferences of the beams reflected at different interfaces. Those Kiessig fringes contain the information on the interface width between the films. The period of the modulation of the intensity is mainly determined by the thickness of the individual films [12]. A detailed fit on the reflectivity curve yields, however, various parameters like film thickness, mean scattering densities, surface roughness, and the interface widths. An example of an x-ray reflectivity curve is shown in Fig. 2. Due to sample preparation the interface between the PS- and PBrS-film has a width due to roughness of 1.1 nm. In the model fits to the experimental data, an error function density profile is assumed. With such a small interface width we are not particularly sensitive about the details of the density profile. During annealing above the glass transition of the two polymers an equilibrium interface width of approximately 2.0 nm is achieved, which does not increase any further with time because of the incompatibility of the materials [14]. The interface width can be roughly explained on the basis of mean field theory where a mean monomer interaction parameter χ is introduced.

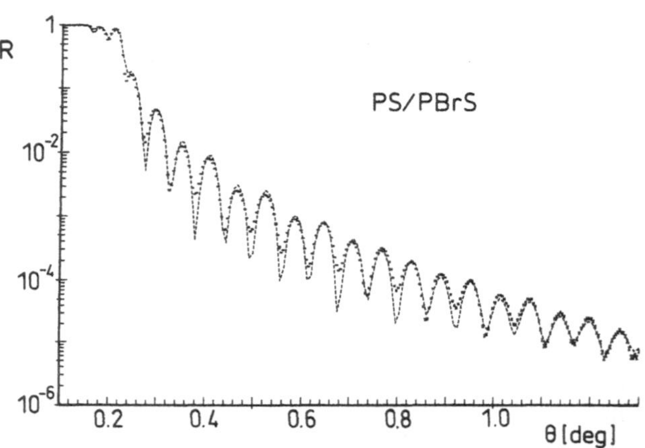

Fig. 2. X-ray reflectivity curve of a PS/PBrS-sample [14]. A thin PBrS-film is deposited on a PS-film spin-cast on a glass substrate. A model fit (dashed line) is also shown which is based on the following parameters: d_1 = 45.0 nm, d_2 = 37.8 nm, σ_0 = 0.6 nm, σ_1 = 1.3 nm, σ_2 = 0.4 nm. Parameters are explained in Fig. 1. The x-ray wavelength is 0.154 nm

For the investigation of initial stages of interdiffusion of PS(H)/PS(D) with time, neutron reflectometry experiments have recently been performed. Annealing procedure, measurement conditions and data treatment are described in [15]. Reflectivity curves at different annealing times are compared in

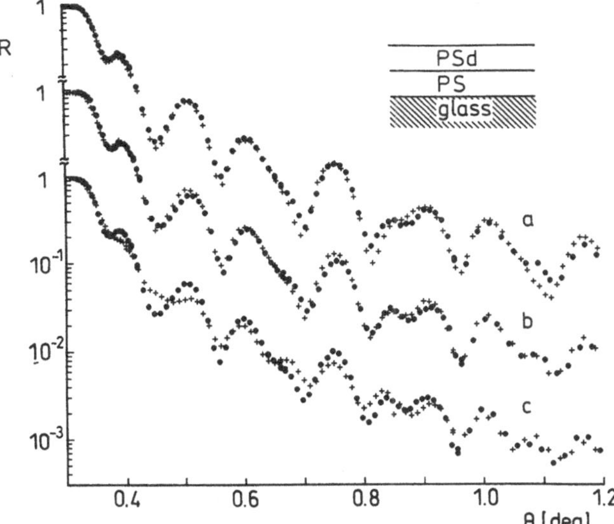

Fig. 3. Neutron reflectometry data of a PS(D) on PS(H) on glass bilayer sample [15]. Different stages of annealing at 120°C are shown: a) unannealed (+) and 2 min (●); b) 2 min (+) and 910 min (●); c) 910 min (●) and 11750 min (+). Changes in different angular ranges at different annealing times are evident. The neutron wavelength is 0.43 nm

Fig. 3. One can clearly distinguish three different regions. At early times a very fast interdiffusion over approximately 2.7 nm is observed, while at annealing times between 2 and 910 min, at 120°C, hardly any change in the neutron reflectivity curve is observed. At still later times, we again observe interdiffusion which is reflected in a change of the reflectivity curve, in particular, also at smaller angles. Since the angle is reversely proportional to a characteristic distance, we can conclude that now the intermixing takes place at a larger distance scale as compared to the earlier stages, where the changes in the reflectivity curve take place at larger angles corresponding to small interdiffusion distances.

This picture is confirmed by simple model fits to the reflectivity curves, assuming an error function profile for the scattering density variation at the interface [15]. These fits are, however, not perfect and a modified density profile based on two error functions is used [16]; it contains a long tail and steep central region and explains the data much better. By a detailed analysis of the neutron reflectivity data, one can thus also obtain information on the shape of the scattering density profile at the interface. This yields additional information on the nature of initial stages of interdiffusion.

One should in general, however, note that the reflectivity technique does not give unique information on the density profile, since the phase information is not present in measured reflectivity curves. Investigations with other supplementary techniques are thus very helpful for the interpretation of data, also if the depth resolution of those techniques is only of the order of 10 nm or larger.

The techniques of x-ray and neutron reflectometry can thus be applied to the investigation of polymer interfaces to yield information with very good accuracy on interface width and density profile. In particular, neutron reflectometry offers many possibilities for the investigation of polymer materials, since a favorable contrast can be achieved by deuteration of components. In many practical cases there may be no other technique available which is capable to also investigate buried interfaces with subnanometer resolution.

Acknowledgeents

Helpful discussions and contributions of Drs. G. Reiter, S. Hüttenbach, and M. Foster are greatfully acknowledged.

References

1. Jones RAL, Kramer EJ, Rafailovich MH, Sokolov J, Schwarz SA (1989) Phys Rev Lett 62:280
2. Coulon G, Russell TP, Deline VR, Green PF (1989) Macromolecules 22:2581
3. Chaturvedi KK, Steiner U, Zak O, Kransch G, Schatz G, Klein J (1990) Appl Phys Lett 56:628
4. Klein J, Briscoe BJ (1976) Polymer 17:481
5. Foster M, Stamm M, Reiter G, Hüttenbach S (1990) Vacuum 41:1441
6. Toney MF, Thompson C (1990) J Chem Phys 92:3781
7. Stamm M, Majkrzak CF (1987) ACS Pol. Prepr. 28(2):18
8. Fernandez ML, Higgins JS, Penfold J, Ward RO, Shackleton C, Walsh D (1988) Polymer 29:1923
9. Russell TP (1990) Mater Sci Reports 5:171
10. Penfold J, Thomas RK (1990) J Phys, Cond Matter 2:1369
11. Stamm M, Adv Pol Sci, Vol. 100, in press
12. Lekner J (1987) Theory of Reflection, Martinus Nijhoff Publ, Dordrecht
13. Stamm M, Reiter G, Hüttenbach S (1989) Physica B 156:564

14. Hüttenbach S, Stamm M, Reiter G, Foster M, Langmuir, in press
15. Stamm M, Hüttenbach S, Reiter G, Springer T (1991) Europhys Lett 14:451
16. Reiter G, Steiner U (1991) J de Phys II 1:659

Author's address:
M. Stamm
Max-Planck-Institut für Polymerforschung
Postfach 3148
6500 Mainz, FRG

Discussion

FLEISCHER:
What is the time-dependence of the profile-broadening in the case of the incompatible polymers?

STAMM:
Because of the small interface width of the investigated system PS/PBrS, we did not investigate the time-dependence of the interfacial broadening. The values given are always obtained in equilibrum when the interface width did not change any more with time at the given temperature.

WÜNSCHE:
Why does the thickness of the interlayer between PS and PBrS not increase above the glass transition?

STAMM:
The increase of thickness below the glass transition temperature T_g(PBrS) is a swelling effect connected with the reduced mobility of PBrS. Above T_g(PBrS) only the temperature dependence of the mean interaction parameter χ between the materials would cause a temperature-dependent broadening. This effect is, however, very small and cannot be resolved within the scope of our experiments.

KREMER:
What assumptions do you need concerning the local profile of the interface between A and B? Might it be that one needs different profiles for compatible and incompatible blends?

STAMM:
In general, we do not need any assumptions concerning the profile of the interface between materials A and B. If the reflectivity data are good enough, one can get a good estimate of the profile, even if it is asymmetric and not of a "simple" form. On the basis of given experimental data this is, however, not always possible, and one usually needs some reasonable model for a fit. One should, of course, assume different profiles between compatible and incompatible materials, in particular if one component is not mobile.

π/A-Isotherms of polymer thin films spread on a Langmuir trough and its discussion on a supramolecular level

E. Köpp, W. Pechhold, and E. Sautter

Abteilung Angewandte Physik, Universität Ulm, Ulm, FRG

If one spreads a small amount of a dilute polymer solution onto a liquid subphase incompatible with the solvent and the polymer, the solvent evaporates and a polymer thin film is formed. As the mobile arm of the Langmuir-gauge moves towards the force transducer, the dispersed pieces of polymer film become condensed and a surface pressure π starts to build up at a certain area A. The main questions to be asked are:

i) what is the area per monomer, when the dense state is reached, and what is the nature of the pressure increase with further compression?
ii) can one explain the π/A-curve — especially if it can be run reversibly several times, as it is the case for BR, IR, NBR — on the basis of a molecular model?
iii) how do the π/A-curves depend on temperature and molecular weight, and how is the electronmicroscopic appearance of the polymer films at different A?

The polymers investigated have not been a subject of much interest in thin films research since 1955 [1]. Their π/A-isotherms do not depend (in a certain range) on the amount of polymer spread on the trough, if plotted versus A/A_0 (A_0 being the area per monomer in a densely packed monolayer).

Some results and its interpretation are:

- the less hydrophobic elastomeres BR, NR, NBR spread nicely reversible on H_2O or H_2O/glycerol mixtures with $A/A_0 \approx 0.5$ at the onset of π. Its π/A-curves can be described by 2-dimensional bundles, superfolded within a monolayer half of which is sheetfolded at $\pi = 0$ and continues to sheetfold during compression. $\pi(A)$ is derived by minimizing a free energy including orientational entropy of segment lines as well as the difference $\delta\sigma$ in surface tensions [2].
- atactic standard polystyrenes — if compressed below the bulk T_g — start increasing π at $A/A_0 \approx 0.13$, suggesting a layer of superfolded 3-dimensional bundles to become meander-folded during compression. A strong relaxation of π in intermittent compression (together with only partial reversibility) seems to indicate a lower T_g for the bundle layer compared to the meander.

References

1. Müller FH (1955) Z Elektrochemie 59:312
2. Pechhold W (1984) Makromol Chem Suppl 6:163

Authors' address:

E. Sautter
Abteilung Angewandte Physik
Universität Ulm
Oberer Eselsberg
7900 Ulm, FRG

Progress in Colloid & Polymer Science Progr Colloid Polym Sci 85:60—65 (1991)

Solid state NMR investigations of polyamide 11 films

P. Holstein, J. Spěváček[1]), D. Geschke, and V. Thiele

University of Leipzig, Department of Physics
[1]) Institute of Macromolecular Chemistry of the Czechoslovak Academy of Science, Prague

Abstract: In this paper, solid-state NMR investigations of polyamide 11 (PA 11) films are presented. This polymer is semicrystalline and can exist in various crystalline polymorphs. Polyamide 11 shows interesting electric properties. The piezo- and pyroelectric effects of this polar polymer are comparable with those of polyvinylidenfluoride (PVDF). Multiple pulse ^{1}H-NMR and ^{13}C CP/MAS NMR measurements are used to investigate the influence of the preparation conditions (thermal, mechanical, electric effects) on the morphology of PA 11. Various preparation procedures provide film material with predominantly α — and γ crystallites, respectively. The ^{13}C NMR spectra and IR measurements support the assumption of the coesistence of various crystalline polymorphs. The chemical shift patterns are different for the various crystalline modifications. Separated ^{13}C NMR spectra are recorded for the crystalline and amorphous parts of the films. Orientation effects due to the mechanical stretching and the poling in electric fields are reflected in the results of the ^{1}H NMR multiple pulse investigations.

Key words: Polyamide 11 (PA 11); solid-state ^{1}H NMR; ^{13}C CP/MAS NMR; electric poling

Introduction

In recent years many publications have been concerned with the investigation of polymeric materials which are polar and therefore show piezo- and pyroelectric properties. The most important polymer of this substance class is polyvinylidenfluoride (PVDF) as it has the most commercial applications [1]. Also, the NMR method was used to obtain information on the nature of the ferro-, piezo-, and pyroelectric properties [2—4].

Normal ^{13}C CP/MAS NMR methods are not convenient for the recording of high-resolution spectra of PVDF, due to the strong additional dipolar interaction caused by the fluorine atoms in this material. Only scanty information has been obtained from ^{13}C CP/MAS NMR spectroscopy for the investigation of piezoelectric active polymers and their internal molecular changes due to external fields (e.g., [5]). There are also other polymers which show a high piezo- and pyroelectricity. Another interesting polymer for applications utilizing the polar properties could be the polar semicrystalline polymer PA 11 [11]. The crystalline phase of PA 11 can exist in two important stable crystalline phases. Both phases, the polar α- and the unpolar γ-phase show remarkable piezoelectric properties. The high values for the piezoelectric constant d_{31} = 7 pC/N and for the pyrocoefficient p = 30 nC/m^2 K are caused by the large dipole moment of the amide group ($m = 1.23 * 10^{-29}$ Cm). The pseudohexagonal γ-unit cells of PA 11 can be poled more easily than the α phase which is based on a triclinic unit cell. In contrast, the α-phase is thermodynamically more stable than the γ-phase [7, 8].

It was shown for PVDF [3, 4] that solid-state proton NMR can be a tool for the investigation of the influence of mechanical stretching and/or high electric poling fields. The ^{1}H NMR methods are also convenient to obtain interesting results for stretched and electrically prepared PA 11 films [9].

In this paper, we report on our experiences from proton NMR investigations with PA 11. In addition, we investigated this polymer by means of high-resolution, solid-state ^{13}C CP/MAS NMR (CP — cross polarization, MAS — magic angle spinning) in order to correlate the results of both methods.

Experimental

The ^{13}C NMR data were obtained from a Bruker CXP 200 spectrometer operating at 50.3 MHz for carbons. The spectra were recorded at room temperature. Typical conditions were 1.5 ms mixing time for the cross polarization, 4 μs for the 90° pulse, and 3 s for the recycle time. The samples were spun with a rotor frequency of about 4.0 kHz. The chemical shift data are referenced to the external shift of glycine. For the MAS-rotor preparation the films were cut into many small pieces, so that in the rotor no preferred macroscopic orientation would occur due to the film pressing or stretching procedures. The solid state ^{1}H NMR measurements were performed using a commercial FT-spectrometer [3]. For various reasons the multipulse group MW4 [10] has been used. The chosen pulse distance parameter τ was 6 μs. The IR spectra were recorded with a Perkin-Elmer 580 B spectrometer. We have investigated polyamide 11 in film form. The material was melt-pressed at T = 200°C for 15 min. For the first type of samples the film sheets were cooled to room temperature with a rate of 2 K/min to favor the crystallization of the α form. Quenching in ice water provides predominantly the γ phase. The oriented samples were produced by zone drawing to a stretching ratio of 3 … 4. Higher stretching temperatures (100°C) induce a higher amount of the polar α phase. After drawing, the film thickness was 30 μm. For the poling of the stretched films aluminium electrodes were evaporated on the film surfaces. The poling was carried out with the thermal electrete poling method. In order to guarantee large polarization effects the film material was poled for the ^{1}H NMR investigations at a field strength of 50 MV/m for 15 min and at a temperature of 75°C.

Results and discussion

Proton solid state NMR investigations were carried out on PA 11 films to detect the influence of the preparation on morphology and orientation effects. The most important preparation procedures for our ^{1}H NMR investigations are as described by Thiele [9]. It was shown that the ^{1}H NMR is convenient to detect relaxation phenomena and orientation effects due to the mechanical preparation (drawing) and/or electrical treatment with the method of the thermal electrete poling. Since PA 11 is semicrystalline, multicomponent magnetization decays were observed in the experiments conducted (free induction decay, multiple-pulse method MW4). In order to investigate separately the effect of the stretching and electric poling, two different types of sections were cut from the films [4], allowing angle variations between the static magnetic field B_0 and the stretching direction (angle γ) and between B_0 and the normal of the film plane (poling direction, angle ϕ). Figure 1 shows an example of the results for the influences of drawing and poling, respectively.

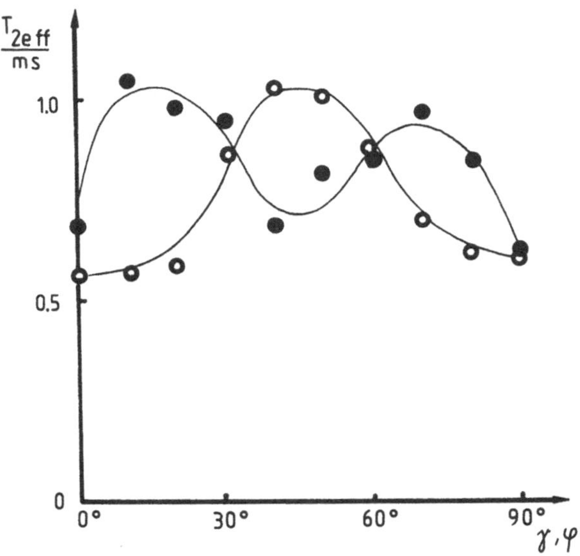

Fig. 1. Angular dependencies of the relaxation time T_{2eff}, ○ angle γ: stretching effect, ● angle ϕ: influence of the poling field, measuring temperature T = 173 K (for explanation of the angles, see text)

The relaxation time T_{2eff} denotes the time in which the magnetization is decayed on 1/e of the start magnetization in the MW4 experiment. Orientation parameters P_2 can be estimated from the angular dependencies of T_{2eff} [4]. For the drawing a typical maximum is seen at the angle γ = 45°. This can be interpreted as a preferred alignment of the main chains in the stretching direction. Before poling, T_{2eff} has the same value for all angles ϕ. After poling, a more complicated angular dependence of ϕ can be detected due to the orientation of the electric dipoles in PA 11 under the influence of the poling field. Line decompositions of these biexponential magnetization decays provide the same values for the crystalline and the amorphous content, respectively, as do the ^{13}C NMR experimental and computational line resolving methods. Some aspects of the ^{1}H NMR investigations are comparable with those on PVDF [3]. PVDF can be poled easily under certain conditions. For PVDF the polarization obtained due to the poling is stable for a long time at normal ambient temperatures. For PA 11 the sample must be cooled after poling in order to investigate the field-induced changes in the films, since poling-induced dipole reorientations are not stable over longer times at ambient temperatures.

The preparation procedures used provide films containing no pure α and γ crystallites; therefore, infrared investigations were carried out. The spectra are different for the annealed (α-phase) and quenched films (γ-phase). This definitely indicates the existence of various types of polymorphs [6, 11]. For stretched PA 11 films, no distinction between α and γ pressed sheets in the IR spectra are seen. Figure 2 shows the IR spectra; the increase in the intensity of the band at 680 cm^{-1} gives an indication for a higher content of α crystallites in the film. But, without knowledge of extinction coefficients or other reference data, we could not estimate an accurate α/γ ratio in the PA 11 films.

For PA 6, the resonance lines were assigned by using the 2D ^{13}C NMR technique [12]. Solid-state ^{13}C CP/MAS NMR spectra of polyamide 6 provide nearly the same peak positions [13]. We have assigned the peaks in the solid-state ^{13}C CP/MAS spectra of polyamide 11 in analogy to PA 6. The carbonyl resonance appears at 173 ppm. In the range of 39 ppm the C_1 peak can be observed. The sign of the carbons and a PA 11 monomer unit are indicated in Fig. 3. The next peak position could be related to the C_{10} carbon. Then appear some additional and little resolved peaks of the other aliphatic carbon atoms. The carbon resonances C_2 ... C_{10} have the same position for all types of samples independent of the preparation procedure; only the peak intensity depends on the preparation conditions, however, further investigations are needed,

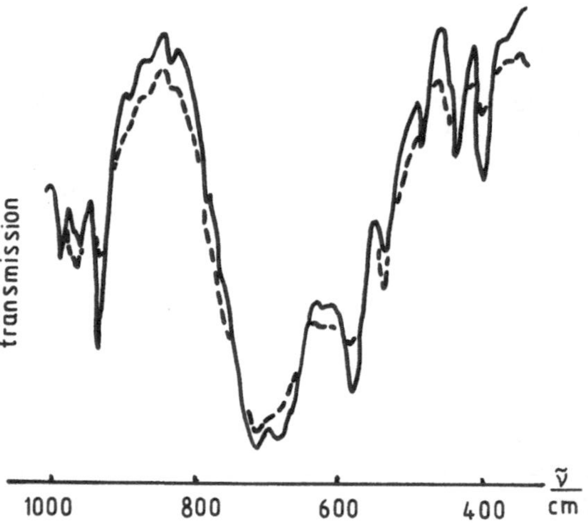

Fig. 2. Infrared spectra of PA 11 films containing preferred α (solid line) and γ (dashed line) crystallites in the wavenumber range 400 cm^{-1} to 1000 cm^{-1}, respectively; $\tilde{\nu}$ denotes the wavenumber

because the peaks are not as well resolved as in the case ^{13}C CP/MAS NMR spectra of PA 6. One reason is the overlapping of more resonances due to the greater number of aliphatic carbons, another reason could be the influences of a distribution of the chemical shifts in the amorphous regions. It is taken into account that the degree of crystallinity is relatively small in this material. T_1 differences were not considered for an optimal cross polarization time.

The most interesting information comes from the C_1 resonance peak. This peak is influenced in a characteristic manner by the existence of the various crystalline polymorphs in PA 11. In Fig. 4 the ^{13}C CP/MAS NMR spectra of a preferred α crystallites containing PA 11 films are presented.

A second peak caused by a C_1 carbon appears at 39 ppm (Fig. 4a). However, the amorphous phase and γ crystallites are responsible for this resonance. The infrared investigations have shown that no pure α or γ phase material occurs under our preparation conditions. This argument is also supported by the separate measurements of the crystalline phase (with a CP T_1 Torchia sequence) and of the amorphous part (^{13}C MAS NMR without CP). Figure 3 shows the chemical shift pattern of PA 11 films containing more γ crystallites. Only one C_1 peak is to be seen at 39 ppm. An analogous observation was made for PA 6 [14]. The

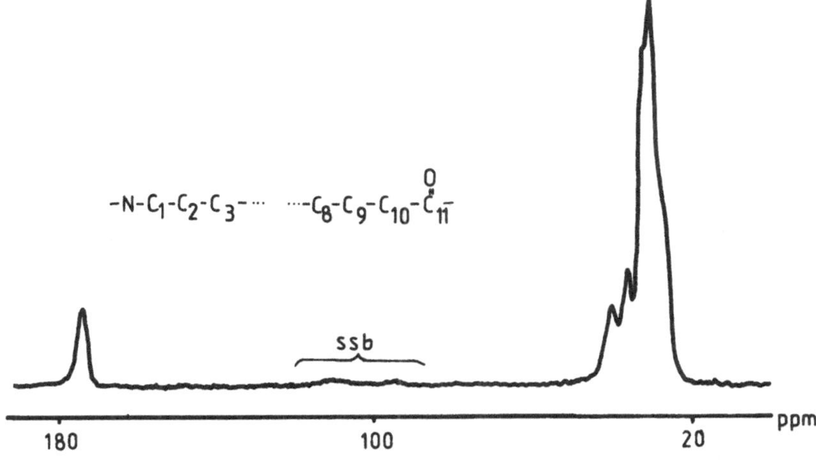

Fig. 3. ^{13}C CP/MAS NMR spectrum of PA 11 prepared with a procedure that induces more γ phase crystallites, over a whole chemical shift range; crosspolarization time t_{CP} = 1.5 ms; ssb are the sidebands; sample rotation frequency; 4 kHz

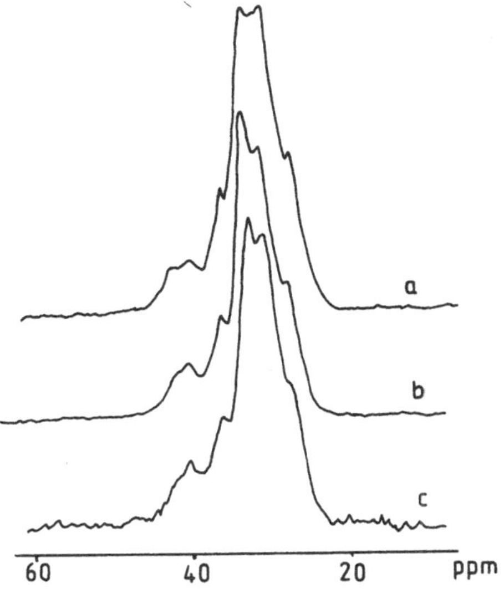

Fig. 4. ^{13}C CP/MAS NMR spectra of polyamide 11, shift range between 0 ppm and 60 ppm, crosspolarization time t_{CP} = 1.5 ms.
a) PA 11 prepared with a procedure inducing α phase;
b) the same sample as in Fig. 4c, but annealed at 102 °C for 80 min;
c) stretched PA 11 α film material (drawing temperature 100 °C, drawing ratio 3)

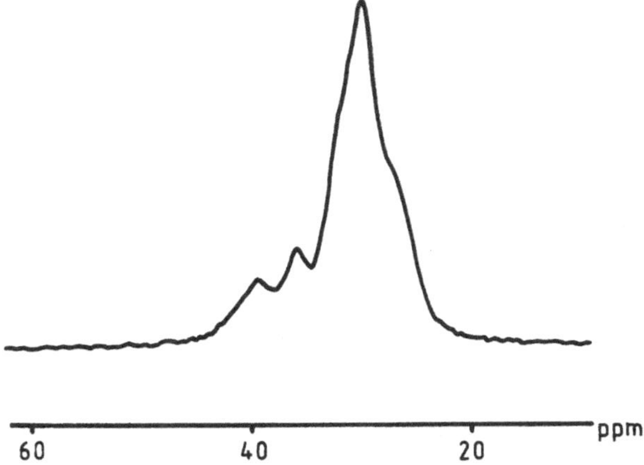

Fig. 5. ^{13}C MAS NMR spectrum of PA 11 without crosspolarization; α and γ film material provide identical spectra of their mobile regions

C_1 shift differences between the various polymorphs can be discussed with the aid of the γ gauche effect. However, more detailed investigations are necessary to interpret this phenomenon. This fact can be used to differentiate the various polymorphs in the polyamide 11 films, and to provide useful information for interpretation of the solid-state ^{1}H NMR results. The success of poling procedures depends strongly on the ratio of the crystalline polymorphs [7]. The ^{13}C NMR spectra of the amorphous regions were recorded with magic-angle spinning, but without crosspolarization. This method preferred the magnetization transfer of the more mobile phase in semicrystalline polymers [14]; the result is shown in Fig. 5. For all types of samples the chemical shift patterns are identically.

This is correct for films containing α and γ crystallites, respectively, and also for the stretched films with a more intermediate ratio of this two polymorphs. This fact could be important for the

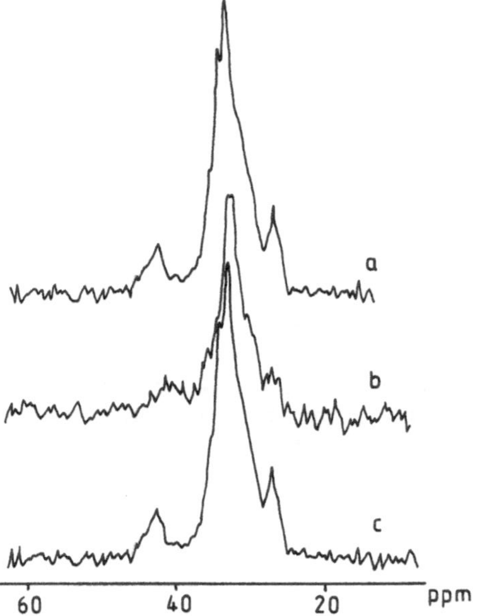

Fig. 6. ^{13}C CP/MAS NMR spectra of polyamide 11 films measured with a Torchia sequence; τ is the pulse distance in the Torchia sequence.
a) PA 11 film containing more α crystallites, τ = 10 s;
b) PA 11 film containing more γ crystallites, τ = 5 s;
c) stretched PA 11 films containing α crystallites, before annealing, τ = 10 s

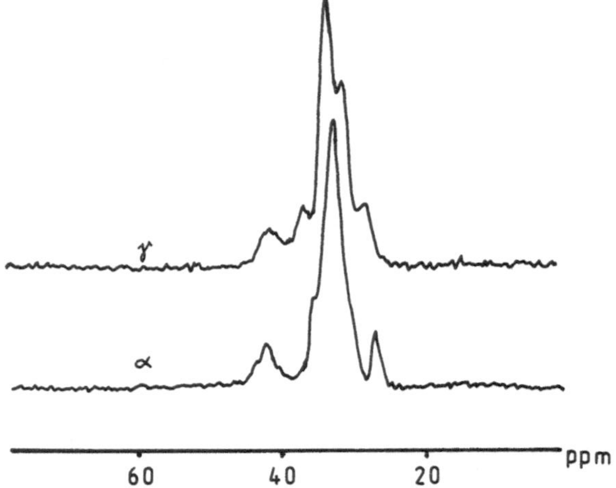

Fig. 7. Spectra obtained from the normalized ^{13}C CP/MAS and ^{13}C MAS NMR spectra for the α and the γ melt pressed film types

discussion of poling effects. The solid state ^1H NMR investigations provide indications that a part of the electric polarization process occurs in the amorphous regions of the polyamide 11 films [9]. It was attempted to obtain an isolated spectrum of the more immobile parts of PA 11. First, short contact time measurements were carried out. We obtained better results with the Torchia T_1-CP/MAS sequence [15]. The differences in the T_1 relaxation times can be used by means of this method for detecting separated spectra of the crystalline and amorphous regions. The spectra are shown in Fig. 6.

As for the ^{13}C CP/MAS NMR spectra, the difference is seen (about 3 ppm) between the C_1 α and the C_1 γ resonance discussed above. The α-C_1 peak is located at 42 ppm (Fig. 6a). The film containing γ crystallites has the C_1 resonance peak at 39 ppm (Fig. 6b). The resonance at 42 ppm is absent. This supports the assumption that this C_1 peak is also related with the amorphous region.

The spectra in Fig. 7 were obtained by digital treatment of the ^{13}C NMR spectra of the semicrystalline samples.

From the normalized spectrum of the whole semicrystalline sample, the normalized spectrum of the amorphous part of the sample (multiplied by the factor f) was substracted. The factor f corresponds to the amount of the amorphous phase. The best fit factors for the predominantly α-crystallite-containing material were f = 0.7 and f = 0.8 for the γ-crystallite-enriched material. We thus conclude that the crystallinity degrees are 30% and 20% for these two preparation types. This result agreed with the results from the proton NMR. The line-shape decompositions of the broad-line spectra and of the biexponential magnetization decay from multiple pulse methods provide the same results for the mobile (amorphous) and the immobile (crystalline) parts of the spectra. Further, Figs. 6a, 6b, and 7 show a good agreement between the calculated spectra of the crystalline phases and the measured crystalline signals.

In addition, the influences of mechanical and thermal treatment on the spectral behavior have been investigated since certain orientation and crystallite phase conditions are necessary for a successful electrical preparation procedure in the electrete sense. For instance, good poling results can be achieved for stretched PA 11 films containing a high amount of the γ crystalline phase.

A stretched film was produced from sheets containing preferred α crystallites. The ^{13}CP/MAS spectrum and the spectrum obtained by using of

Torchia sequence before the cross polarization are shown in Figs. 4c and 5c. In Fig. 4b the ^{13}C CP/MAS spectrum is presented for this film after annealing. The aim of the annealing at 102°C for 80 min was to achieve a higher γ content in the film. Relaxations of molecular orientation and modification changings are produced by thermal effects. A reversible α-γ transition occurs if the temperature is increased over such temperatures. The change of the crystallite type occurs for PA 11 at temperatures of about 100°C [16, 17]. It is known from the T_{2eff} temperature dependencies [3] that an additional relaxation appears in this temperature region. This is reflected in a high-temperature minimum in the temperature curve. The transformation of the modification can be see in the IR spectra measured immediately after the annealing and the cooling to ambient temperatures. But, the obtained amount of γ crystallites is small. However, the conversion of the modification is not stable over the time. This can be concluded from the ^{13}C CP/MAS NMR spectra. It seems that the modification transformation is fully relaxed. The recording of the NMR spectra was started about 30 h after the annealing. Also, the ^{13}C CP/MAS NMR measurements using a Torchia sequence show the same chemical shift patterns as before the annealing procedure. The T_1-CP/MAS spectra measured after and before annealing are identical (Fig. 5c). The reconversion of the crystalline type occurs during some hours. Further investigations are needed to improve our preparation procedures to achieve higher and more stable amount of the γ crystalline phase and to find correlations between the film preparation and the spectroscopic data.

Acknowledgements

We are indebted to Dr. P. Schmidt from Prague (Institute of Macromolecular Chemistry) for recording the IR spectra and useful discussions.

References

1. Gerhardt-Multhaupt R, Gross B, Sessler GM (1987) In: Sessler GM (ed) Electrets. Springer pp 383—431
2. Douglass DC, McBrierty VJ, Wang TT (1982) J Chem Phys 77:5826—5835
3. Geschke D, Holstein P (1989) Progr Colloid & Polymer Sci 80:71—77
4. Geschke D, Holstein P (1990) Makromol Chem Macromol Symp (1990) 34:205—211
5. Jo YS, Muramaya J, Inoue Y, Chujo R, Tasaka S, Miyata S (1988) J Polym Sci Phys Ed 26:463—466
6. Wu G, Yano O, Soen T (1986) Polymer J 18:51—61
7. Scheinbaum JI, Mathur SC, Newman BA (1986) J Polym Sci: Part B: Polym Phys 24:1791—1803
8. Newman BA, Sham TP, Pae KD (1977) J Appl Phys 48:4092—4098
9. Thiele V, Geschke D (1990) Acta Polymerica 41:550—551
10. Mansfield P, Ware D (1966) Phys Lett 23:412—422
11. Hummel D, Scholl F IR Atlas of Polymers, Interscience, New York 1969 Vol 1, Part 2
12. Ketels H, Schellekens R, Beulen J, Van Der Velden G (1988) Polymer Commun 29:189
13. Ketels H, Van de Wen L, Aerdts A, Van der Velden G (1989) Polym Commun 30:80—83
14. Kubo K, Yamanobe T, Komoto T, Ando i, Shiibashi T (1989) J Polym Sci: Part B: Polym Phys 27:929—937
15. Torchia DA (1978) J Magn Res 30:613
16. Mathias LJ, Powell DG, Autran JP, Porter RS (1990) Macromolecules 23:963—967

Authors' address:

Dr. P. Holstein
Universität Leipzig
Sektion Physik
Linnéstr. 5
O-7010 Leipzig, FRG

Dipole reorientation in polymeric layers investigated by photo-induced current

D. Althausen, P. Wünsche, and E. Peškova*)

Department of Physics, University Leipzig
*) Institute of Macromolecular Chemistry, Czechoslovak Academy of Sciences

Abstract: Photo-induced currents were studied in poly(n-vinylcarbazole) (PVCa) without external electric field. A polarization of dipoles exists due to corona charging in PVCa. This polarization relaxes to thermal equilibrium after generation of quasi-free charges by a short flash and the transit of these charges through the photoconductor. The relaxation of dipoles causes an opposite anomalous current. For interpretation, a model is developed which takes into consideration a depolarization by dipole reorientation due to the decay of the electric field caused by the charge carrier transit. The measured time-dependence of the anomalous photocurrent can be interpreted by the numerical application of this model.

Key Words: Photo-induced current; depolarization; poly(n-vinylcarbazole)

Introduction

Photo-induced currents in photoconductors are mainly determined by charge generation and transport. The charge generation process was omitted in the following, because it takes place in a very short time compared to the charge transport and the additionally considered dipole relaxation.

Current-voltage characteristics are taken in steady-state for investigation of charge transport. The characteristics are described for space-charge-limited conduction in darkness as follows: an ohmic region first occurs with increasing voltage, followed by a trap-modulated current, a trap-filled limited current and, finally, a trap-free region. Of course, this is only a general picture [1, 2].

Another way to investigate charge transport is a short excitation, for instance by light, of charged dielectrics, followed by measurements of the time development of charge density. To investigate the residual voltage a sandwich arrangement with two electrodes on both sides of the sample is used [3,4]. The electrodes were short-circuited after a voltage application and then opened. The voltage can be measured isothermally. Additionally, such investigations can be completed with thermally stimulated current measurements.

Photo-induced discharge measurements are the obvious choice to investigate charge transport in photoconductors. The surface potential is recorded as well for long and for relatively short (ms) illuminations. It was found, for instance, that the perturbation of the charge transport for Se/PVCa is caused by a space charge which is generated by high photon flux. The mobility of holes in PVCa was estimated as $4 \cdot 10^{-7}$ cm²/Vs [5].

Another method used in this context is the time-of-flight (TOF) method [6]. A steady-state field is applied to the sample and the current is recorded after a short light flash. Measurements of the time-dependence of the photocurrent over 10 decades in time are known [6,7]. In polymers [9], as in a-Si:H [8], the time dependence of the current frequently obeys

$$I \propto t^{-1+\alpha_1} \quad t < t_T$$
$$I \propto t^{-1-\alpha_2} \quad t > t_T \,, \tag{1}$$

where I is the current, t the time, α_i characteristic parameters of the photoconductor, and t_T the transit time. The advantage of TOF measurements is the ideal situation of only few moving charge carriers relative to the capacitor charge of the arrangement combined with a constant external electric field. This permits to neglect influences of space charges or of polarization in the sample.

But, generally, the dielectric volume properties and the structural determined trap distribution in the sample additionally influence the photocurrent effect. Hence, very extensive work is done to investigate this trap structure in polymers by thermally stimulated luminescence (RTL) and currents (TSC) [10—13].

Several successful theories were established to interpret the TOF experiments. The measurements can be interpreted by a thermally activated hopping transport in PVCa [14]. Most hopping processes in polymers can be described by the multiple trapping theory (MT) with successive trapping and thermal releasing of charge carriers during the transit through the dielectric [15, 16]. A trap-rate distribution could be found in the case of PVCa which is nearly flat [7]. The continuous-time random walk model (CTRW) is another way to describe the carrier transport [17]. The distribution of the hopping times is almost continuously in disordered systems due to the manifold possible distances and activation energies of hopping. This distribution could be asymptotically expressed as a function of hopping time [18]. Monte Carlo simulations are carried out (beside MT and CTRW) to describe the currents measured by TOF. So it was possible to ascertain that hopping rates are influenced by the structure of the polymeric matrix and that these rates govern the charge transport [19, 20].

Looking to the literature, it can be pointed out that most of the polymeric photoconductors consist of molecules with strong electric dipoles. In the case of PVCa with sorbed O_2 the dipole moments are in the order of 4...5 D, for instance [13]. The dipoles can form a macroscopically evident polarization according to the applied electric field. The electric field in the photoconductor caused by charges and the polarization will be changed by charge transport. In addition, the polarization is affected by field variations and follows the actual electric field with its own time constant.

No investigations of the influence of polarization on the charge transport are known. The steady external electric field and the small amount of moving charges prevent polarization changes in TOF investigations and photo-induced discharge characteristic (PIDC) measurements are carried out, usually with long illumination which allows the polarization to follow the field changes within the generation and transport time of the charge carriers. Additionally, a high input resistance of the electronic equipment is necessary for PIDC investigations. Hence, these investigations are connected with a high time constant of the experimental arrangement.

We developed an experimental setup to investigate the influence of polarization on charge transport in photoconductors [21]. In our arrangement the samples were polarized by a corona discharge and then pairs of charges were generated by a short laser flash. After that, the charge carrier transport takes place and the change of the surface potential is recorded. Because of the great amount of moving charges the internal field varies widely and the internal polarization relaxes to a new thermal equilibrium. This polarization decay influences the charge transport.

Photo-induced currents recorded with the described experimental set-up will be presented in the following section. Then the theory for the temporal development of the field and the current is described, including dipole reorientation, and the comparison between theory and experimental results will be given in the final section.

Experiment

It is necessary to allow internal field alteration during the charge transport in order to investigate the influence of polarization on charge transport due to field changes. That is why the experimental set-up described in [21] was used.

The photoconductor is charged by a corona discharge. The surface potential amounts to 450 V for PVCa layers of 1-μm thickness in our equipment after a corona voltage of 5 kV during 1 s [22]. A transparent electrode (area about 1 cm^2) is brought over the photoconductor after the corona. A current measured by a preamplifier connected with the electrode can be recorded if the photoconductor is exposed by a short laser flash (wavelength 337.1 nm, energy density about 8 $\mu J/cm^2$, half time 2.5 ns). After the measurement, the photoconductor is discharged by 50 laser flashes, and then the measurement is repeated without corona voltage to get the zero signal. This run iterates about 100 times for accumulation.

The time constant of the arrangement is small enough to record the current with a dwell time of 0.1 μs because of the low input resistance of 59 Ω, the input capacity of about 15 pF, and the measuring capacity of about 1 pF. The error of the current measurement is estimated to be smaller than 30 μA.

Very pure samples of PVCa were prepared in the Institute of Macromolecular Chemistry of the Czechoslovak Academy of Sciences. A 8% PVCa solution in toluene-cyclohexane (4:1) was cast on supports. The samples were dried in air. The thickness of PVCa layers was measured with a surfometer SF200.

We investigated the photo-induced currents for different thicknesses of the PVCa layers to decide whether the measured unusual part of the current is determined by surface or volume effects. All recorded currents show a similar behavior. After an initial drop the current reaches zero and changes its polarity. A minimum is traversed before this anomalous current becomes undetectable. In Fig. 1 photo-induced currents are presented measured on PVCa layers with different thicknesses. A larger thickness causes a later alteration of the current polarity. Also, the absolute minimum value becomes smaller with increasing thickness.

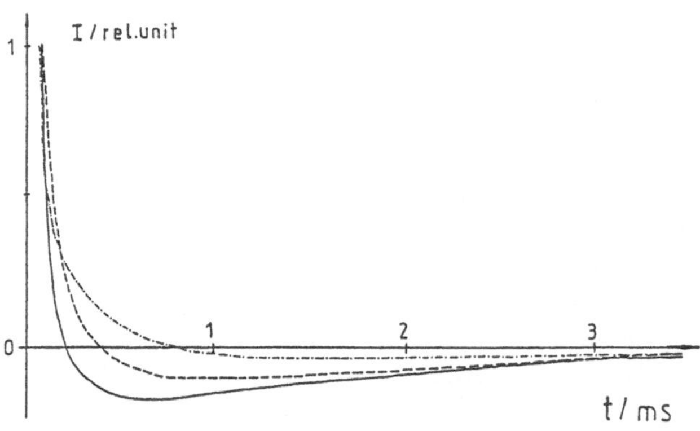

Fig. 1. Photo-induced current for different thicknesses of PVCa layers (corona voltage 5 kV, corona duration 1 s, temperature 30 °C). —— = 1 μm; ---- = 2 μm; —·—·— = 6.8 μm

Theory

Our theoretical model consists of a capacitor with a dielectric polymeric layer of thickness L. One electrode of the capacitor is directly connected with the lower side of the polymer layer and the other one has the distance s relative to the polymeric photoconductor. With Kirchhoff's second law one gets

$$sE_{air} + R_i I = \int_0^L E(x)dx = U_{eff} , \tag{2}$$

where E_{air} is the electric field in the air gap, R_i the input resistance of the preamplifier, and E the field within the photoconductor. The voltage drop in R_i can be neglected since even the light conductivity of the photoconductor is smaller than $10^{-13}\ \Omega^{-1}\ cm^{-1}$ [23, 24]. A current in the input resistance R_i of the preamplifier is observed due to changes of E_{air} at the distance s as a result of the internal field changing. Hence the measured current I_{mea} is proportional to dE_{air}/dt

$$I_{mea} = -\varepsilon_0 A \frac{dE_{air}}{dt} , \tag{3}$$

where ε_0 is the dielectric constant of vacuum and A the area of irradiation. The electric field is connected with the charge density ρ and the polarization P by Poisson's equation. The reduction of the calculation to one dimension is possible due to the very big ratio of each length of the considered area A relative to the thickness of the photoconductor. Integrating Poisson's equation over two spatial dimensions yields

$$\begin{aligned} l(x) &= \int_A \rho da = \frac{dq(x)}{dx} \\ &= A \frac{\delta}{\delta x} (\varepsilon_0 E(x) + P(x)) , \end{aligned} \tag{4}$$

where $l(x)$ is the one-dimensional charge density (charge per length) and $q(x)$ the charge. The time variation of the integral of Eq. (4) over the thickness of the photoconductor gives the measuring current

$$\begin{aligned} I_{mea} = &-\frac{A}{s} \int_0^L \left[\frac{1}{A} \frac{\delta q(t,x)}{\delta t} - \frac{\delta P(t,x)}{\delta t} \right] dx \\ &- \frac{AL}{s} \left[\varepsilon_0 \frac{\delta E(t,x=0)}{\delta t} - \frac{1}{A} \right. \\ &\left. \cdot \frac{\delta q(t,x=0)}{\delta t} + \frac{\delta P(t,x=0)}{\delta t} \right] . \end{aligned} \tag{5}$$

The boundary condition at the surface $x = 0$ is

$$\varepsilon_0 E_{air}(t) + \varepsilon_0 E(t, x = 0) + P(t, x = 0) = \frac{1}{A} q(t, x = 0) \,, \tag{6}$$

since the dielectric displacement is changed by the surface charge $q(t, x = 0)/A$. Using Eqs. (3), (5), and (6), one gets

$$I_{mea} = \frac{A}{s\left(1 + \frac{L}{s}\right)} \int_0^L \left[\frac{1}{A}\frac{\delta q(t,x)}{\delta t} - \frac{\delta P(t,x)}{\delta t}\right] dx \,. \tag{7}$$

Now the evaluation of $\delta q/\delta t$ and $\delta P/\delta t$ is necessary. The charge conservation law within the photoconductor is

$$\frac{\delta \rho}{\delta t} + \frac{\delta j_L}{\delta x} = 0 \,. \tag{8}$$

The relationship between the net charge current j_L and the electric field is

$$j_L = \mu \rho E \,, \tag{9}$$

where μ is the mobility of the charge carriers. The temporal development of the charge density can be evaluated by

$$l(t,x) = l(0,x) - \mu \int_0^t \frac{\delta E(t',x) l(t',x)}{\delta x} dt' \,. \tag{10}$$

Every charge carrier will be omitted at the counter electrode, so we use the following boundary condition for $x = L$:

$$l(t, L) = 0 \,. \tag{11}$$

Now the polarization must be regarded. The polarization depends on time according to

$$\frac{\delta P}{\delta t} = \frac{(P_s - P)}{\tau} \,. \tag{12}$$

τ is the mean relaxation time of the dipoles. The steady-state polarization P_s is connected with the electric field by

$$P_s = aE \,. \tag{13}$$

One gets with Eqs. (12) and (13):

$$P(t,x) = \left[P(0,x) + \int_0^t \frac{a}{\tau} E(t',x) \cdot \exp\left(\frac{t'}{\tau}\right) dt'\right] \exp\left(\frac{-t'}{\tau}\right) \,. \tag{14}$$

No analytical solution for E could be found for the resulting integro-differential-equation (Eqs. (10) and (14) used in (4)). Hence, the temporal change of the surface potential was evaluated by computer.

By using a finite quantity of $\Delta x = L/N$ and $\Delta t = t/Z$ the integral over x of the Poisson's equation supplies the electric field

$$E(t, x = n\Delta x) = E(t, x = 0) + \frac{\Delta x}{2A\varepsilon_0}\left[l(t, x = 0) + 2\sum_{m=1}^{n-1} l(t, m\Delta x) + l(t,x)\right] - \frac{1}{\varepsilon_0}[P(t,x) - P(t, x = 0)] \,. \tag{15}$$

Equations (2) and (6) are used for calculating $E(t, x = 0)$:

$$E(t,x{=}0) = \left[\frac{1}{\left(1 + \frac{\Delta x}{2s}\right)}\right] \cdot \left\{\frac{1}{\varepsilon_0}\left[\frac{q(t,x{=}0)}{A} - P(t, x = 0)\right] - \frac{\Delta x}{2s}\left[2\sum_{m=1}^{N-1} E(t, m\Delta x) + E(t,L)\right]\right\} \tag{16}$$

With Eq. (10) the discrete charge density can be calculated by

$$l(t,x) = l(0,x) - \frac{\mu \Delta t}{2}\left[\frac{d_D}{dx}E(0,x)l(0,x) + 2\sum_{v=1}^{Z-1}\frac{d_D}{dx}E(v\Delta t,x)l(v\Delta t,x) + \frac{d_D}{dx}E(t,x)l(t,x)\right], \quad (17)$$

where the "discrete" gradient d_D/dx is denoted as

$$\frac{d_D X_D(n\Delta x)}{dx} = \begin{cases} \frac{[X_D((n+1)\Delta x) - X_D(n\Delta x)]}{\Delta x} & \text{for } n=0, \\ \frac{[X_D((n+1)\Delta x) - X_D((n-1)\Delta x)]}{2\Delta x} & \text{for } 0<n<N, \\ \frac{X_D(n\Delta x) - X_D((n-1)\Delta x)]}{\Delta x} & \text{for } n=N. \end{cases} \quad (18)$$

The electric field is regarded to be constant during the time Δt for calculating the polarization $P(t,x)$. Thus, with Eq. (14), one gets

$$P(t,x) = \left[P(0,x) + \frac{a}{\tau}\sum_{v=0}^{Z-1}\int_{v\Delta t}^{(v+1)\Delta t} E(t',x) \cdot \exp\left(\frac{t'}{\tau}\right)dt'\right]\exp\left(-\frac{t}{\tau}\right)$$
$$\approx \left\{P(0,x) + a\sum_{v=0}^{Z-1}E(v\Delta t,x) \cdot \left[\exp\left(\frac{(v+1)\Delta t}{\tau}\right) - \exp\left(\frac{v\Delta t}{\tau}\right)\right]\right\}\exp\left(-\frac{t}{\tau}\right). \quad (19)$$

The charge density $l(t,x)$ and the polarization $P(t,x)$ can be evaluated with Eqs. (15)–(19). The electric field $E(t,x)$ is determined both by the charge density and the polarization, and itself influences the charge density and polarization in their time-dependencies.

A surface charge σ arises during corona charging of the photoconductor. The maximum penetration depth of the charge carriers is estimated to be <0.2 μm [25]. Hence, the penetration of the charge carriers during and after corona charging was neglected. The surface charge $\sigma(x = 0)$ produces the electric field $E_C(t = -0,x)$ in the photoconductor $(0 \leqslant x \leqslant L)$. The dipoles are oriented in the field $E_C(-0,x)$ until the laser flash. Hence, $E_C(t = -0,x)$ reduces to

$$E(t = -0,x) = \frac{1}{\varepsilon_r}E_C(t = -0,x) = \frac{\sigma}{\sigma_0\varepsilon_r}, \quad (20)$$

where ε_r is the relative permittivity. Charge carrier pairs are generated by the photons within a short time compared to the measuring time [26]. A part of the generated charge carriers recombines with the surface charge. Another part of the generated charge carriers recombines together. The remaining charges are quasi-free and can be transported to the counter electrode. Since the charge carriers are generated near the surface of the photoconductor by the laser flash, a fast recombination with the corona charges is assumed and disregarded in the calculations. With Lambert's law and charge conservation during the laser flash the charge density can be calculated to be

$$l(+0,x) = \frac{A\alpha\varepsilon_0\varepsilon_r U_{\text{eff},0}}{L[1 - \exp(-\alpha L)]}\exp(-\alpha x), \quad (21)$$

where α is the absorption coefficient. The initial polarization rises up in the field of the corona charges before the laser flash:

$$P(t = -0,x) = aE(t = -0,x) = \frac{aU_{\text{eff},0}}{L}. \quad (22)$$

The initial polarization is assumed to be

$$P(t = +0,x) = P(t = -0,x), \quad (23)$$

since the relaxation time of the dipoles is much greater than the time between laser flash and the start of measurement. Starting with the evaluated initial conditions, the charge density and polarization at the time Δt were calculated. The electric field at the time Δt was estimated using these quantities. Then the charge density and polarization at $2\Delta t$ were computed with the electric field at Δt. To estimate the error of this algorithm, we remark that the changes of the electric field are greater at the beginning than at any following time. This happens due to the faster transport of charges followed by the slower dipole relaxation. The error of the charge density calculation can be estimated by

$$\Delta l(t,x) \leqslant | \, l(\Delta t, x)_{\text{with } E(\Delta t)} - l(\Delta t, x)_{\text{with } E(0)} \, |$$

$$= \frac{\mu \Delta t}{2} \left| \frac{d_D}{dx} E(\Delta t, x) l(\Delta t, x) - \frac{d_D}{dx} E(0,x) l(\Delta t, x) \right| . \qquad (24)$$

$\Delta l(t,x)$ depends on the choosen time interval Δt which can be evaluated with these initial conditions as

$$\Delta t \leqslant \frac{l(0,(N-1))}{\mu \left[\left| l(0,0) \frac{d_D}{dx} E(0,0) \right| + \left| E(0,(N-1)) \right| \left| \frac{d_D}{dx} l(0,0) \right| \right]} , \qquad (25)$$

supposing the error in the order of the lowest amount of the charge density at $x = (N-1)\Delta x$. The gradient of the electric field was estimated to be smaller than twice that of the greatest field at the beginning $E[0,(N-1)]$. The resulting Δt is on the order of 5 ... 50 ns. This is small compared to the dwell time of the measuring equipment and the characteristic times of the considered process. Thus, Eq. (25) was used to determine the optimal Δt. (Note that this Δt depends indirectly on Δx by the gradients.)

In the proposed model any space charge is neglected which may be generated at the barrier between the photoconductor and the counter electrode. A diffusion of charge carriers is possible, but neglectable after the corona charging. Also, in the theoretical model no remaining surface potential is regarded.

Discussion

The electric field $E(t,x)$ was calculated with Eqs. (6), (15), and (16) at the start and in dependence on time (Fig. 2). The initial surface potential is V_0 = 400 V, the mobility μ = 2.9 · 10^{-6} cm^2/Vs, the coupling factor a = 2 · 10^{-13} As/Vcm, the relaxation time τ = 7 ms, the absorption coefficient α = 4.35 · 10^4 cm^{-1}, and the thickness L = 1 µm. After the corona discharge the surface charge σ causes a homogeneous electric field which is reduced due to the stimulated internal polarization. The photons of the laser flash generate charge carrier pairs. Hence, the total charge is conserved in the photoconductor during the flash. The charges additional to corona charges are immediately generated according to Lambert's law, whereby a fast recombination is assumed with the opposite corona charges. Therefore, the field is reduced in the irradiated part of the photoconductor due to lowering of the charge density in this part by remaining polarization. This starting field is the border curve at t = 0 in Fig. 2. The charges are quickly transported through the dielectric, and the depolarization only slowly reduces due to the greater relaxation time of the dipoles. Hence, the field is increasingly determined by the polarization. The electric field is inverted completely if the greater part of the charges has passed the photoconductor (curve t_{14} in Fig. 2). Finally, the field will become zero with relaxation of dipoles. The creases of the curves are caused by the values of the parameters Δt and Δx used during calculation, but do not change the general picture of the time-dependence of the electric field.

The measured current corresponds to the integral of the time derivation of the electric field calculated by Eqs. (2) and (3). Four parameters, the relaxation time τ, the coupling factor a, the mobility μ, and the initial potential V_0 influence the resulting current. For instance, the calculated current is shown for different relaxation times τ in Fig. 3. The disappearance of the current occurs more slowly after the minimum with increasing relaxation time. The

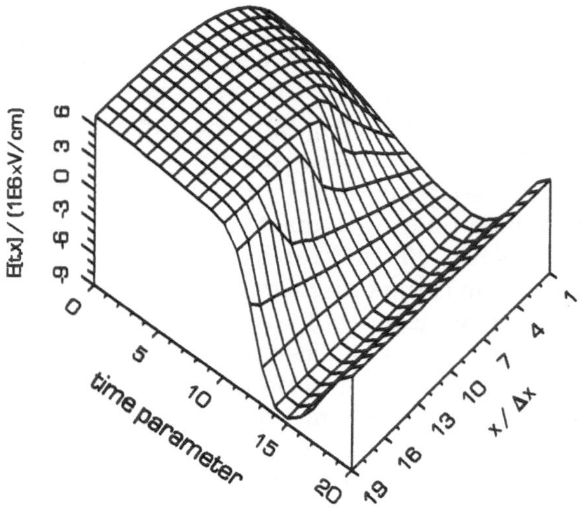

Fig. 2. Calculated electric field $E(t, x)$.
Parameter of the curves is the time t depicted in logarithmic scale. Time steps:
$t_0 = 0$ $t_{14} = 1.29 \cdot 10^{-4}$ s
$t_1 = 1.38 \cdot 10^{-8}$ s $t_{15} = 2.58 \cdot 10^{-4}$ s
$t_2 = 2.76 \cdot 10^{-8}$ s $t_{20} = 8.26 \cdot 10^{-3}$ s
(Additional parameters are given in the text)

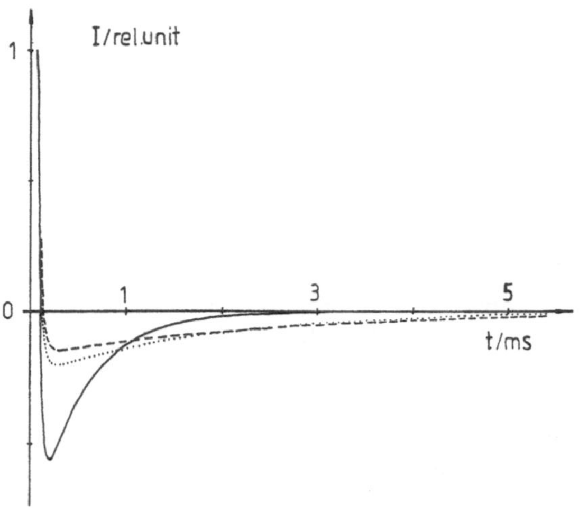

Fig. 3. Calculated currents with various relaxation times. —— =2 ms; ... = 7 ms; ––– = 10 ms. (Physical parameters correspond to those of Fig. 2)

depolarization current increases with decreasing relaxation time (see Eq. (12)). Hence, the absolute value of the minimum is affected. The relaxation times are distributed in PVCa, which can be seen from dielectric investigations (e.g., [13]). But we only used a mean relaxation time in our calculations, because the discussion of the distribution function of the relaxation times is not possible using only our experimental values.

Of course, various velocities of the charge carriers, represented by the mobilities μ, influence the current, too, causing different initial drops of it (Fig. 4). The decay of the current is faster for higher mobilities.

The influence of the thickness on the measured current is depicted in Fig. 1. The prolongation of the time of the crosspoint of the measured current with the abscissa, caused by increasing of the thickness, can be interpreted by a longer transit time of charges passing the photoconductor. This behavior can be represented by the calculation, too.

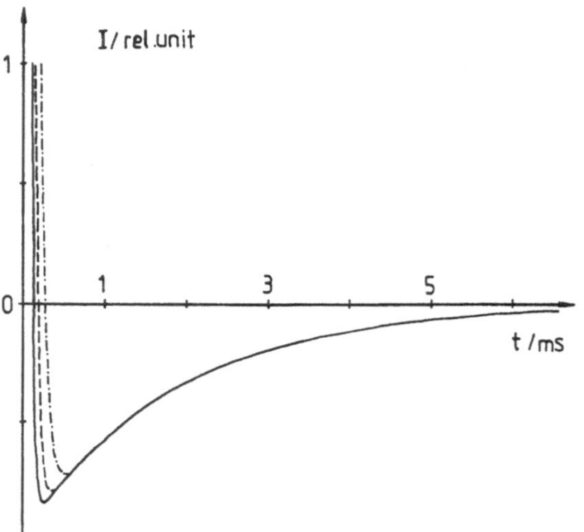

Fig. 4. Calculated currents with various mobilities. —— = $3.1 \cdot 10^{-6}$ cm²/Vs; ––– = $2.0 \cdot 10^{-6}$ cm²/Vs; —·—·— = $1.2 \cdot 10^{-6}$ cm²/Vs. (Physical parameters correspond to those of Fig. 2)

It is concluded that the measured current shows itself to be a volume process because of the thickness dependence of the current and due to the remaining anomalous current with increasing thickness.

The measured curves were fitted with calculated ones, starting with mobilities from literature ($1.2 \cdot 10^{-6}$ cm²/Vs) [14] (Fig. 5). The parameters V_0, a, τ, and μ, which give the best fit of the measured currents, were determined as follows. The initial potential V_0 was measured and amounts 400 V [22]. The

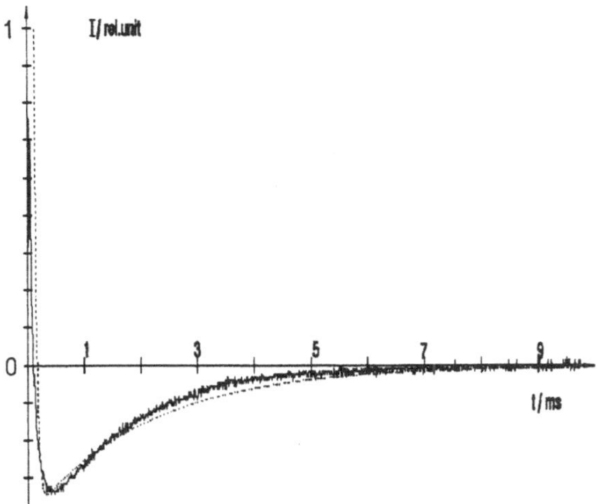

Fig. 5. Calculated (...) and measured (——) currents for a 1-μm layer of PVCa

fitted coupling factor a was $2 \cdot 10^{-13}$ As/Vcm. This corresponds to a dielectric constant of $\varepsilon_r^{fit} = 3.24$ which is near the dielectric constant $\varepsilon_r = 3.0$ given in the literature [27]. The relaxation time of the polarization was determined by the theoretical fit to be 7 ms. The order of magnitude of this value can be verified with the current decay after the minimum of the anomalous part of the measured current. The mobility of the charge carrier amounts to $2.9 \cdot 10^{-6}$ cm²/Vs. This resulting mobility of the fit is greater than that obtained by TOF measurements [14]. To discuss this fact the field dependence of the mobility is taken into consideration [14]. According to the experimental conditions of our investigations, the electric field strongly changes during the charge transport. With the results of the TOF method for PVCa [14] the field dependence of the mobility of charge carriers has been estimated for zero field and an electric field strength of 400 V/(1 μm) as $1.8 \cdot 10^{-8}$ cm²/Vs and $4 \cdot 10^{-4}$ cm²/Vs, respectively. Hence the mobility calculated by our fit ($2.9 \cdot 10^{-6}$ cm²/Vs) is regarded as a mean value of the field-dependent mobility varying during our experiment.

The remaining deviations between the measured and calculated curves in Fig. 5 can be attributed to the values of Δt and Δx (used during the calculations), together with neglecting the distribution of relaxation times and of the field-dependence of the mobility.

The theories of charge transport mentioned in the introduction use the electric field to create a preferential direction for charge carrier motion. In his basic formula for changing of the charge density in the conduction band, Schmidlin used a term equal to the gradient of the product of charge density, mobility, and electric field [16]. In the CTRW model, Scher and Montroll defined transition probabilities corresponding to nearest-neighbor hopping on a simple cubic lattice with a bias for jumps in the preferred direction due to the electric field [17]. There is no electric-field changing process in these equations, thus, the electric field must be considered as constant. Various charge carrier currents are presented in Fig. 6 calculated using different coupling factors between polarization and electric field. As can be seen, the coupling factor has an influence on the charge carrier current, however, different coupling factors cause various depolarizations. Due to this process the electric field is changed and modifies the charge transport. Therefore, it is concluded from our theoretical approach and experimental results that field changes are possible and have to be taken into consideration during investigations of photocurrents.

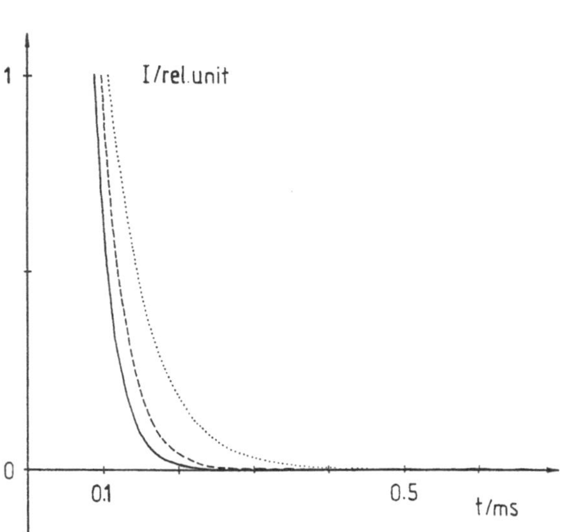

Fig. 6. Charge carrier current for various coupling factors. —— = $3.0 \cdot 10^{-13}$ As/Vcm; – – – = $2.4 \cdot 10^{-13}$ As/Vcm; ... = $1.5 \cdot 10^{-13}$ As/Vcm. (Physical parameters correspond to those of Fig. 2)

References

1. Wintle HJ (1983) In: Bartnikas R, Eichhorn RM (eds) (1983) Engineering dielectrics volume IIA, Electrical properties of solid insulating materials, molecular structure and electrical behavior, American Society for Testing and Materials, Philadelphia
2. Zor M, Hogarth CA (1987) Phys Stat Sol (a) 99:513
3. Kaneko F, Kobayashi S (1985) Electrical Engineering in Japan 105:18
4. Kitani I, Arii K (1986) Jpn J Appl Phys 25:1332
5. Mort J, Chen I, Emerald RL, Sharp Jh (1972) J Appl Phys 43:2285
6. Spear WE (1957) Proc Phys Soc London B70:669
7. Muller-Horsche E, Haarer D, Scher H (1987) Phys Rev B 35:1273
8. Overhof H, Thomas P (1989) Electronic Transport in Hydrogenated Amorphous Semiconductors, Springer Verlag, Berlin-Heidelberg-New York
9. Yoshino K, IEEE Trans EI-21:999
10. van Turnhout J (1975) Thermally Stimulated Discharge of Polymer Electrets, Elsevier, Amsterdam
11. Fleming RJ (1990) Radiat Phys Chem 36:59
12. Akuetey G, Hirsch J (1989) J Phys D 22:174
13. Pochan JM, Hinman DF, Nash R (1975) J Appl Phys 46:4115
14. Gill WD (1972) J Appl Phys 43:5033
15. Ieda M (1984) IEEE Trans EI-19:162
16. Schmidlin FW (1977) Phys Rev B 16:2362
17. Scher H, Montroll EW (1975) Phys Rev B 12:2455
18. Bos FC, Guion T, Burland DM, Phys Rev B 39:12633
19. Stolzenburg F, Ries B, Bässler H (1987) Ber Bunsenges Phys Chem 91:853
20. Bässler H (1989) 9. Tagung Polymerphysik, Potsdam
21. Wünsche P, Althausen D (1990) Macromol Chem, Macromol Symp 37:27
22. Koy U (1990) unpublished results, University Leipzig
23. Hoegl H, Süs O, Neugebauer W (1957) Ger Pat 1111935, 1068115
24. Shattuch MD, Vahtra U (1969) US Pat 3484237
25. Sessler GM (ed) (1987) Electrets, Springer Verlag, Berlin-Heidelberg-New York
26. see for example Duke CB (1982) In: Kirk-Othmer Encyclopedia of chemical technology, John Wiley & Sons, Inc
27. Peasson JM, Stolka M (1981) Poly(N-Vinylcarbazole), Gordon and Breach Sci Publ, New York

Authors' address:

D. Althausen
Universität Leipzig
Sektion Physik
Linnèstr. 5
O-7010 Leipzig, FRG

Mechanism of the formation of polyaniline films at transparent conducting electrodes and characteristics of their optical and electrochemical properties

K.-H. Heckner and A. Uhlig

Institut für Physikalische Chemie, Humboldt-Universität zu Berlin

The formation of PANI by succesive cyclic voltammetry or by potentiostatic technique at 0.6 V SCE shows typical pecularities of the phase formation (induction period for the nucleation and a following fast growing of the phase). By use of current transients and transmissions vs. time curves, charge ballances and measurements of the deuterium isotope effects was found details of the kinetics an mechanism and the role of oligomers during the formation of PANI by anodic induced polymerization of aniline.

For the investigations of electrochemical transformations within the PANI films was used spectroelectrochemical and current transients in the different redox states (—300 mV, +300 mV at +600 mV) and the change of the transmission spectrum in the visible range. The anioninsertion and — exsertion and the protonation/deprotonation processes in the different redox states of PANI are measured by radiotracertechnique ($^{36}Cl^-$ and ^{3}H). From the results was supposed a mechanism of the electrochromic reactions and the different forms of PANI which take place in the electrochemical transformation. On the basis of our results some problems are discussed of application of PANI to pratical electrochromic displays devices.

Authors' address:

K.-H. Heckner
Institut für Physikalische Chemie
Humboldt-Universität zu Berlin
Bunsenstraße 1
O-1080 Berlin

Defect morphology in preoriented systems with constrained perturbations

F. Plümer

Department of Polymer Physics, Institute of Physics, University of Leipzig, FRG

Abstract: A computer simulation model has been extended to discuss a special crystallization event in thin oriented polymer films, subjected to a current shearing at simultaneous quenching. Supposing in the model at first an isotrope defect distribution in a highly extensional strained system, the development of this distribution as a result of a spinodal decomposition process has been computed. Especially the assumption of a shift of chains against each other in the direction of orientation affecting the process of defect migration and collection has been included in the model. The computed time-dependent structure factor $S(k, t)$ seems to be able to indicate influences of chain movements at the development of the defect morphology. By that it should be possible to state introducing as perturbation a chain shift the defect collection will be strong affected. This is compared to a sudden rupture of the formation of fibrillar structures changing slightly the preparation conditions.

Key words: Defect morphology; orientation; polymer films; spinodal decomposition; Monte-Carlo simulation

Introduction

Preparation procedures, including shear forces for manufacturing thin semicrystalline polymer films from the melt are often characterized by an orientation controlled structure development. Accordingly, the extent of chain orientation within this melt defines decisively the picture obtained of the structure, especially in very thin polymer films crystallizing from highly extensional strained melts.

Experimental results, reported by Petermann and Gohil [1] about such systems, exhibit fibrils having density modulations along their chain axis. Investigations of Zachariades [2] have been concerned with crystallization of thin polymeric layers, cooling down from the melt between cylindrical plates rotating against each other. If the melt has been exposed simultaneously a pressure of 20 MPa fibrillar structures could be observed. Our own experiments [3] carried out under comparable conditions by an injection-molding procedure show similar results, as do preparations of Melior and Hirte [4] (but with a marked change of the crystallization event, as shown by SALS-, IR- and NMR-investigations [3]). Only small deviations of the preparation conditions, especially a reduction of the applied pressure, lead to a nearly complete loss of the expected fibrillar structures.

Meakin [5] proposes to compare such a crystallization event in highly oriented polymer systems with a computer-assisted simulation model, taking as a basic the concept of spinodal defect decomposition of a two-phasic system, as also discussed in [6]. He assumes regions like fibrils composed of extremely extended chains formed during the shearing of the melt. Defects such as kinks, jogs or chain ends, at first randomly distributed throughout the system and undergoing a phase separation under annealing treatments, determine the temporal and spatial starting position for the following structural development. Accepting this simple model of defect morphology to investigate the observed sudden structural deviations as the aim of this work, the following question arises: if the proposed model is able to become extended by introducing a further perturbation term, which causes a nonstationarity of the original chain arrangement after quenching the extended melt (especially a successive shift of the chains representative of a persistent shear) is the later computed picture of the defect morphology able to give hints about such a sudden change of the structural development, especially about a reduction or an incomplete collection of defects?

The model

The proposed two-dimensional Monte-Carlo-simulation-model [5, 7, 8] with uniaxial kinectics of the defect migration process has been constructed by a basic set of points representing the defects, randomly distributed at a two-dimensional 32 * 64 square lattice and wrapped on a torus in order to realize boundary conditions. Not distinction will be made between the different kinds of defects. Assuming all chains are first oriented in the x-direction, this initial lattice then represents a random configuration of defects projected at a oriented background belonging to an infinite temperature of this system. The mutual influence of the defects arises in a first approximation as a result of a power-law interaction between them with

$$H = -J \sum_{l,m} n_l \cdot n_m / r_{l,m}^p \, . \tag{1}$$

If there is a defect on site l, then $n_l = 1$; otherwise, $n_l = 0$. The range of interaction is limited by p. Other assumptions leading especially to cooperative processes as discussed in the crystallization theories for polymer systems [6] are not taken into account within this model.

The system will be quenched to a temperature T well below the critical temperature T_c and the interactions overcome the averaging by the heat bath [9].

Depending on the angle between the connection vector and the direction of orientation of the chain axis, interacting defects influence the hopping probability for the defect migration along the chain axis, which will computed by

$$P_{i,j} = \tau^{-1} \exp(-\beta U_{ij})/(1 + \exp(-\beta U_{ij})) \, , \tag{2}$$

where $\beta = (k_B T)^{-1}$ and U_{ij} is the change in the energy of the subsystem which would result from the interchange [8]; τ represents the relaxation time.

After a certain number of hoppings per defect under quenching conditions, a separation of defects in a banded structure perpendicular to the chain orientation will be expected [5]. The resulting lattice occupation after any hoppings per defect (HD) can be examined by computing the time-dependent structure function $S(k,t)$, the Fourier-transform of the pair correlation function $G(\vec{r},t)$:

$$G(\vec{r},t) = N^{-1} \sum [c(\vec{r}_i,t)c(\vec{r}_i + \vec{r},t) - \bar{c}^2] \, ,$$

$$\bar{c} = N^{-1} \sum c_i \, , \text{ the arranged defect concentration;} \tag{3}$$

$$S(k,t) = \sum \exp(ik\vec{r}) G(\vec{r},t) \, ;$$

$$k = (2/64)\pi \cdot \mu \, , \quad \mu = 0, 1, \ldots 63 \, . \tag{4}$$

To complete the model the two kinds of perturbation should be introduced. The first one, a sinosidal perturbation transferred with starting the quench realizes an imprinted structure at the defect system by addition of a sinosidal perturbation amplitude to the energy term in (2), whereas a second kind of perturbation should simulate a "softening" after the quench of the, at first, oriented and frozen chains, realized by a weak and successive shift of the chains against each other in the x-direction, meaning a persistent shear within the quenched system.

These computations are carried out at first only for the component in the direction of chain orientation (to economize computation time), suggested, moreover, by the basic idea of this model as a uniaxial kinetics one.

Results

Figures 1 and 2 show the development of a randomly distributed defect pattern after quenching to a temperature $T = 0.85\ T_c$ and 60 hoppings per defect without perturbations. The generation of a banded defect concentration perpendicular to the horizontal assumed chain axis is seen. The starting random distribution was found to be without any correlation, as shown in Fig. 3, representing the $S(k,t)$ for these cases. In the early stage of the decomposition process after 5 or 10 hoppings per defect the formation of a peak at $k = 0.8$ can be observed; later the growth and the shift to a smaller k is in a good agreement with [5]. All $S(k)$-values are plotted in the same arbitrary units (y-axis). Figure 4 shows the temperature depend-

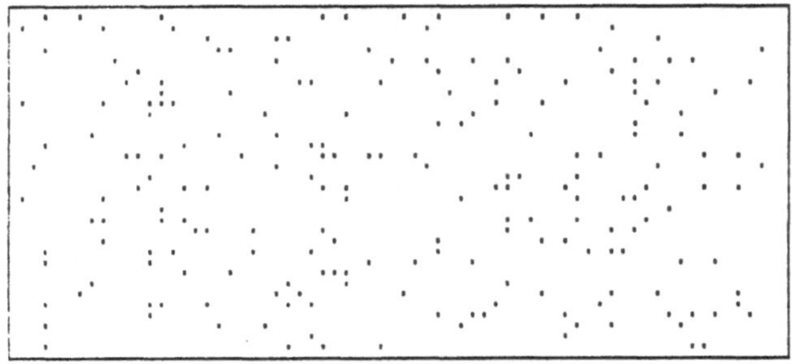

Fig. 1. 10% random defect pattern 32 * 64; orientation assumed horizontal

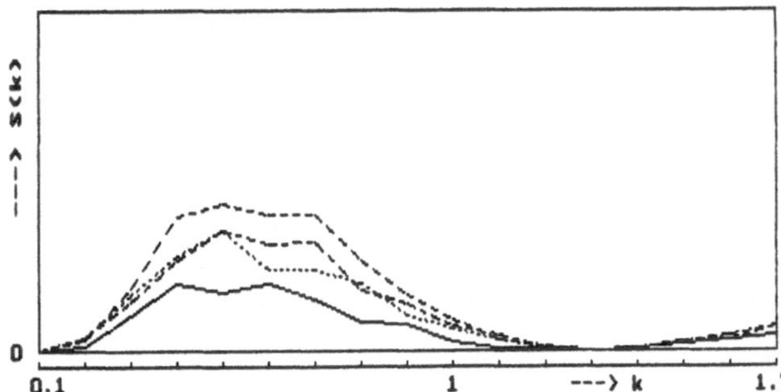

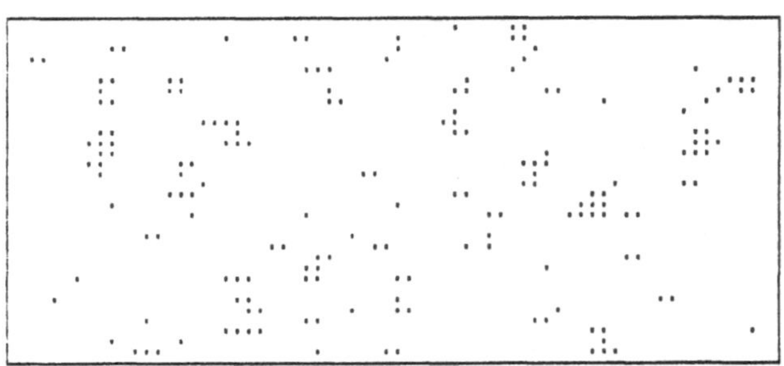

Fig. 2. Defect pattern after 60 hoppings per defect at $T = 0.85\ T_c$

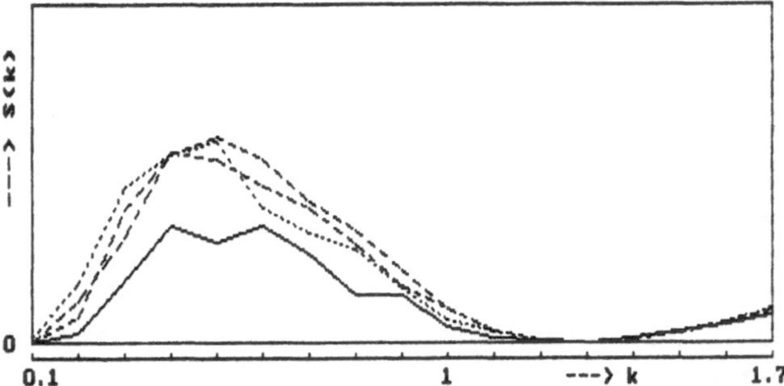

Fig. 4. Temperature dependence of $S(k, t)$: —— 0.5, ... 0.7, – – – 0.85, —·—·— 1 T_c; above: after 40 HD; bottom: after 60 HD

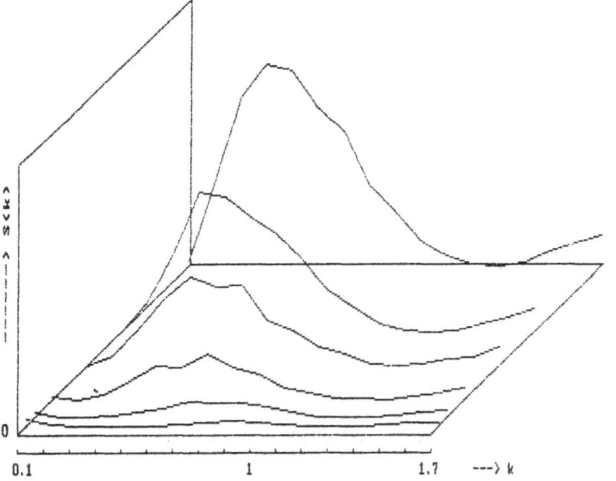

Fig. 3. Structure factor $S(k, t)$ after 0, 5, 10, 20, 40, 60 HD's at $T = 0.85\ T_c$ without perturbation

ence of the defect migration for two hopping situations (HD's) also executed at an unperturbed lattice. If a small sinosidal perturbation is introduced, indicating a preordered melt and characterized by the wave-vector k, an additional peak at this wave-vector position was found, as shown in Fig. 5. Note that after like 20 HD's, the growth of the perturbation peak is finished, whereas the peak induced by the defect interaction more and more increases and is quite unaffected by the perturbation. The perturbation dependences of $S(k)$ at two temperatures are shown in Fig. 6 (other constant parameters are noted). A shift of the perturbation peak is found as a function of k, but simultaneously, we see also the nearly unaffected behavior of the shear interaction peak. On the other hand, the perturbation peak increases the more the wave-vectors for perturbation and interaction peak are in agreement. This can be extended to other quenching temperatures, as shown in Fig. 7. It is interesting to note the opposite behavior of the two temperature dependences: the left power peak increases, the right one representing the sinosidal perturbation decreases with temperature.

Summarizing the effect of this kind of perturbation, the conclusion is that in all cases of good ac-

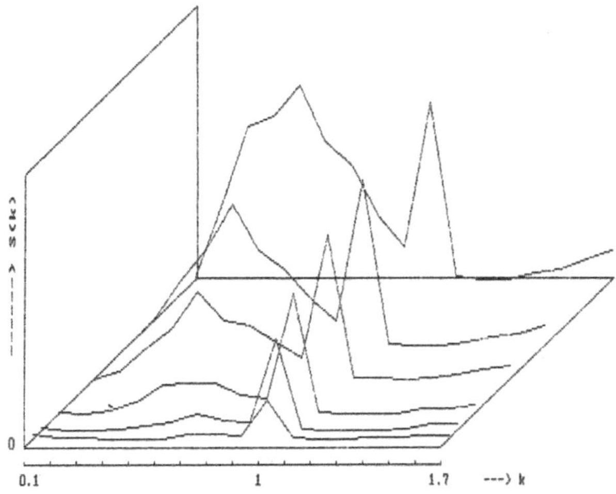

Fig. 5. $S(k, t)$ after 0, 5, 10, 20, 40, 60 HD's at $T = 0.85\ T_c$ and sinosidal perturbation with $k = 1$

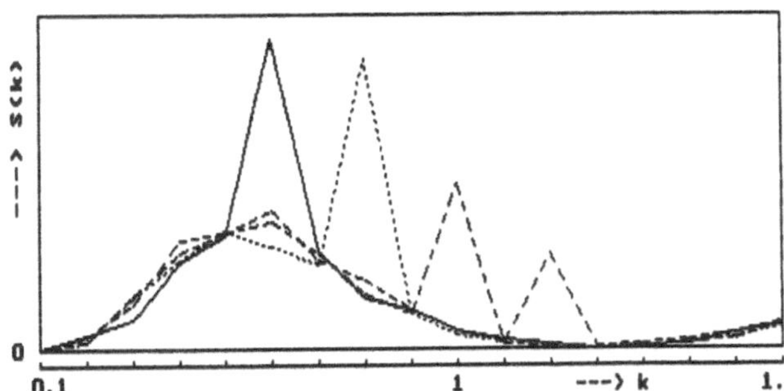

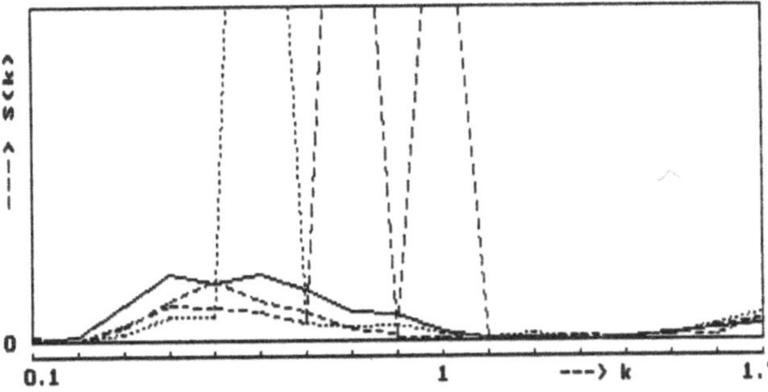

Fig. 6. Wave-vector dependence of $S(k, t)$ for 2 temperatures:
above: $T = T_c$; —— 0.6, ... 0.8, ——— 1, —·—·— 1.2 k
bottom: $T = 0.5\ T_c$; —— 0, ... 0.6 ——— 0.8 —·—·— 1 k

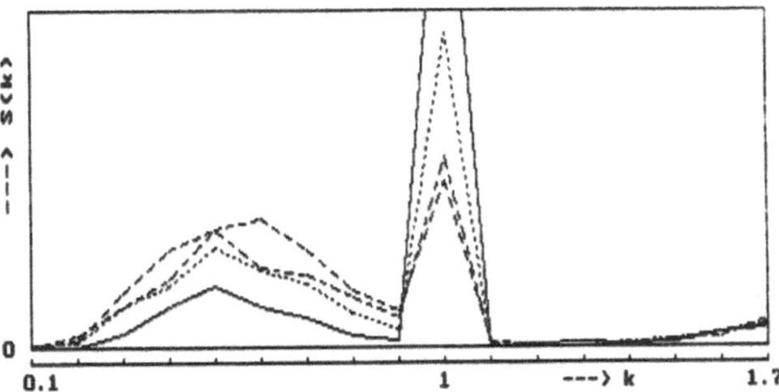

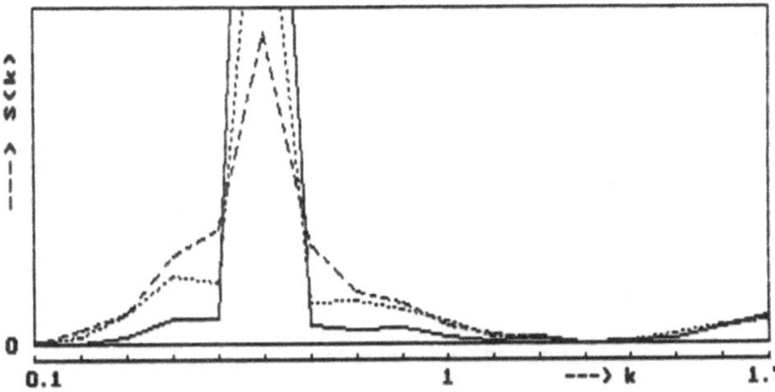

Fig. 7. Temperature dependence of $S(k, t)$ for 2 wave-vectors:
above: $k = 1$; —— 0.5, ... 0.7, ——— 0.85, —·—·— 1 T_c
bottom: $k = 0.6$; —— 0.5, ... 0.7, ——— 1 T_c

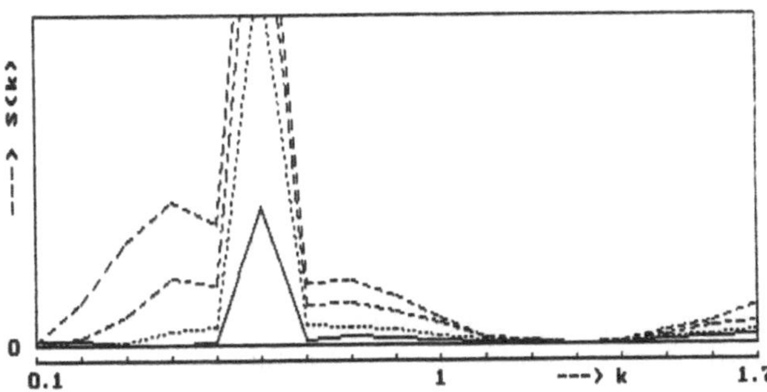

Fig. 8. $S(k, t)$ for $T = 0.7\ T_c$ and $k = 0.6$ after 10, 20, 40, 60 HD's without shift

cordance in k between the sinosidal perturbation and the power law interaction the overall correlation increases over the unperturbed level; in all others the transferred initial structure remains constant or will be increasingly destroyed.

If one now introduces in the model the second kind of perturbation (shear) by a stepwise shift,

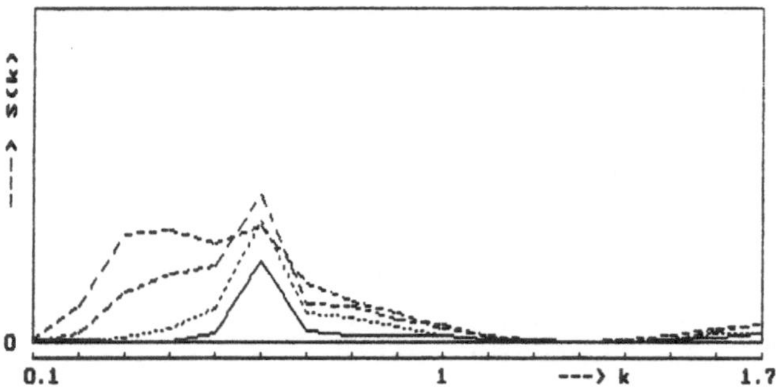

Fig. 9. As Fig. 8 with a shift of 1 step per 10 HD's

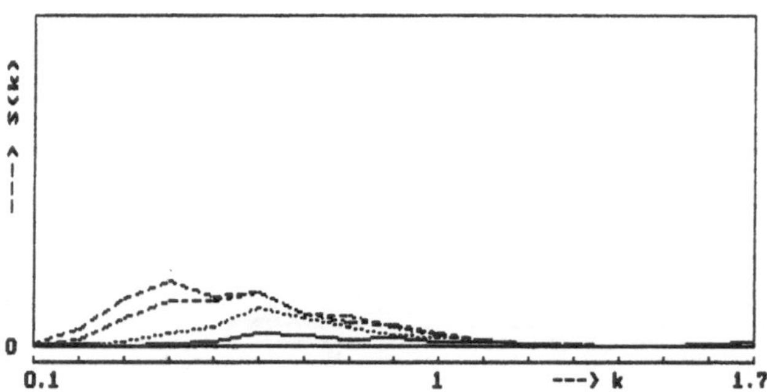

Fig. 10. As Fig. 8 with a shift of 1 step per 1 HD

with every successive chain against the former one at one position with respect to the postulated boundary conditions (and in terms of the time defining hoppings per defect), the following behavior can be observed.

Superposing the defect pattern additionally with a small shift of one site of a chain against the adjoining one per 10 HD's as a realization of a weak shear, both the sinosidal perturbation term with $k = 0.6$ and the temperature of quenching are constant; the result is a marked reduction of the peak at $k = 0.6$, as shown by relating this behavior in Figs. 8 and 9.

Shifting the chains one step per 1 HD (Fig. 10) in a tenfold stronger way, one can observe an important decrease of the separation effect. The peak concerned is reduced by one order of magnitude. An equlibrium between the increasing correlation caused by the power law interaction and a sinosidal perturbation and — on the other hand — the decreasing correlation caused by the chain shifting can be observed as a function of the chosen parameters. Therefore, at first the qualitative conclusion should be allowed that the possibility to realize a small shift of the defined chains against each other, especially permitted by reducing the pressure within the chain system under shear, is able to prevent the formation of a coined and banded structure within the defect pattern. This indicates, in the sense of the introduction, a hindrance of a fibrillar structure formation in the prepared polymer layers. Other arguments, especially the problems of the real heat transfer throughout the system after the suddenly assumed quenching which provokes in the simulation model, for example, a position-exchange of the chains will be discussed later.

References

1. Petermann J, Gohil RM (1979) J Mater Sci 14:2260—2264
2. Zachariades AE (1984) J Appl Polym Sci 29:867—875
3. Plümer F (1987) In: Abstr VI, 31. IUPAC-Symp Macro 1987 Merseburg, p 27
   Kramer M (1987) Diplomarbeit, Leipzig, unpublished
4. Melior JP, Hirte R (1987) In: Abstr IV, 31. IUPAC-Symp Macro 1987 Merseburg, p 162
5. Meakin P, Scalapino DJ (1985) J Polym Sci, Phys Ed 23:179—189
6. Fischer EW, Stamm M, Fan G, Zietz R (1989) Am Chem Soc: Polym Preprints 30/2:291—292
   Baumgärtner A, Heermann DW (1986) Polymer 27:1777—1780
7. Binder K (1979) Monte Carlo Methods in Statistical Physics, Springer, New York
   Binder K (1983) J Chem Phys 79:6387—6407
   Pistoor N, Binder K (1988) Colloid Polym Sci 266:132—140
8. Bortz AB, Kalos MH, Lebowitz JL, Zendejas MA (1974) Phys Rev B 10:535—541
   Marro J, Bortz AB, Kalos MH, Lebowitz JL (1975) Phys Rev B 12:2000—2011
9. Onsager L (1944) Phys Rev 65:117
   Metropolis N, Rosenbluth AW, Rosenbluth MN, Teller AH, Teller E (1953) J Chem Phys 21:1087—1092

Author's address:

Dr. Friedrich Plümer
Universität Leipzig
Sektion Physik, WB Polymerphysik
Linnestr. 5
O-7010 Leipzig, FRG

Submicron particles with thin polymer shells

W.-D. Hergeth, U.-J. Steinau, H.-J. Bittrich, K. Schmutzler, and S. Wartewig

Department of Physics, "Carl Schorlemmer" Technical University of Leuna-Merseburg, Merseburg, FRG

Abstract: The general conditions for the formation of thin polymeric shells on nm-sized particles in aqueous emulsion polymerization systems are discussed. The core-shell structure of the particles was verified by small-angle scattering (x-ray, neutrons) and electron microscopy. Differential scanning calorimetry, nuclear magnetic resonance, and vibrational spectroscopy were applied to determine thickness and chemical composition of the interfacial layer between core and shell polymer.

Key words: Polymer; polymer composites; emulsion polymerization; core-shell structure; interfacial layer

I. Introduction

Emulsion polymerization is a well-known technique for producing dispersions with a well defined structure of polymer particles. One of these well defined structures is the core-shell structure. Here are some of the advantages and applications of dispersions with polymeric core-shell particles:

i) The polymers with core-shell structure are perfect model systems in polymer physics because of their regular distribution of components and the simple spherical geometry of the system. Polymer blends and composites can thus be obtained for studying problems in materials sciences.

ii) Investigations of core-shell structures and of their formation afford a better overall understanding of emulsion polymerization and especially of the topochemistry of the reaction during the period of particle growth.

iii) It is possible to modify the interfacial properties of polymer particles in the aqueous phase by producing definite core-shell structures. Film properties of such dispersions can thus be improved. Hence, such dispersions can be applied in industry for producing adhesives, coatings or paints.

The aim of this paper is to discuss, from the point of view of physics, some aspects of the preparation and characterization of core-shell colloids. In the first part, experimental conditions for the formation of a definite core-shell structure will be outlined. The detection of this type of internal particle structure will be summarized in the second section. The third section deals with the determination of thickness and chemical composition of the interfacial layer between core and shell material for both polymeric and inorganic cores using various techniques. The investigations were carried out using poly (vinyl acetate) (PVAC) as one component of the system.

II. Experimental

The polymerizations were carried out as two-stage reactions in an all-glass reaction vessel, constantly stirred (250 r.p.m.), at a constant temperature (343 K). Seed latexes (monomer I/polymer I) were prepared applying recipes and techniques known from the literature. In the second step, the seed latex (polymer I) was added to the reaction vessel and heated, and stirred continuously, to the reaction temperature. Then the aqueous solution of initiator ($7 \cdot 10^{-3}$ mol/dm^3 $K_2S_2O_8$ referred to the whole mixture and preheated to the reaction temperature) was added, and the continuous dosage of monomer II was started simultaneously under monomer-starved conditions. The procedure has been described in detail elsewhere [1, 2].

The monomer conversion was determined gravimetrically. The particle size was obtained by conventional light scattering and electron microscopy (Tesla BS 613 microscope).

Experimental equipment and calculating procedures of small-angle neutron scattering (SANS) are described elsewhere [1, 3, 4]. The intensities recorded at a wavelength of $\lambda = 1.085$ Å were corrected for background and desmeared.

The small-angle X-ray scattering (SAXS) profile was measured using a Kratky-camera. The experimental counts ($\lambda = 1.54$ Å) were corrected for scattering in the empty equipment and smoothed using cubic spline functions [5].

Measurements of the specific heat capacity were carried out using a DSC-2C (Perkin-Elmer) at scanning rates of 20 K/min. All parameters were calculated from second heating runs.

^{1}H NMR relaxation measurements of transversal magnetization were carried out at a resonance frequency of 88 MHz using a SXP 4-100 spectrometer (Bruker). IR spectra were recorded on a Specord M 80 spectrometer (Zeiss, Jena).

III. Formation of shells

The seeded polymerizations using water soluble initiator were carried out applying a seed latex particle number which was high enough to prevent the formation of separate particles during the second stage of the reaction. This minimal seed latex particle number N_I can be estimated following a proposition by Schmutzler, extensively discussed in [1, 6, 7]:

$$N_I^2 \gg \frac{3\,k_{11}}{8\pi k_{12}^2 r_0^3 \rho_\tau} \cdot \frac{dM_{II}}{dt}\,, \qquad (1)$$

where k_{11} and k_{12} describe the coagulation rate constants between primary particles and latex particles, respectively. r_0 and ρ_τ are the radius and the density of primary particles and dM_{II}/dt is the monomer conversion per unit time of the second stage polymerization.

According to Eq. (1), an additional experimental condition can be derived which is necessary for shell formation. dM_{II}/dt is the monomer conversion in the aqueous phase, which is proportional to $[M]_w$ (i.e., the monomer concentration in water). This has to be as low as possible in order to fulfill (1). Hence, the second stage of the emulsion polymerization should be carried out under monomer starvation. The monomer-II addition rate should be lower than the corresponding monomer consumption rate during emulsion homopolymerization.

Moreover, monomer starvation promotes shell formation on given seeds in a further way because of the resulting lower monomer concentration within the particle. This leads to a comparably high viscosity of the particles' polymeric material (e.g., with respect to batch polymerization condition) which reduces the diffusion length of the radicals from water phase into the particles [2].

Under these circumstances, the shell formation is very inhomogeneous if incompatible polymers were used as polymer I and II. On the micrograph (Fig. 1), one can see small "islands" of polystyrene (PS) on a PVAC core latex particle. These islands grow by continued polymerization according to diameter $\sim C_{PS}^{1/3}$, where C_{PS} is the weight fraction of PS (Table 1). The number of islands per particle remains nearly constant at 5...7. These islands touch each other at a shell/core mass ratio of 0.30 to 0.35 and then build up closed but nonspherical shells at a ratio of 0.40 to 0.45. The particles are spherical at a mass ratio equal to or greater than 0.70.

In the case of PVAC — poly (methyl methacrylate) (PMMA) core-shell polymers, the diameters of the PMMA islands are only one-third to one-half as large as the corresponding diameters of the PS islands on PVAC cores at the same shell/core mass ratio. The formation of closed PMMA shells occurs at shell/core mass ratios greater than 0.30.

In the opposite succession of polymers (PS-PMMA, PS-PVAC), we observed a more homogeneous growth of the particles. Surface inhomogeneities owing to polymer II are more numerous and reach maximum diameter in the range of 10 to 15 nm [2].

These experimental findings can be explained by two facts: i) a spreading effect of polymer II on polymer I, and ii) a selective absorption of monomers and/or selective adsorption of oligomers.

i) The Gibbs Free Energy of a core-shell particle ΔG_{CS} contains mixing and interfacial terms

$$\Delta G_{CS} = \sum_{i<k} (\Delta G_{MIX})_{ik} + \sum_{i<k} (\Delta G_{INT})_{ik}\,. \qquad (2)$$

During two-stage emulsion polymerization, the formation of polymer II takes place in a manner such that ΔG_{CS} will either be reduced or the increase of ΔG_{CS} will be as slight as possible. In the case of incompatible polymers, the mixing term is positive $(\Delta G_{MIX})_{ik} > 0$. Hence, the polymers are separated from each other. However, this phase separation takes place within the particles because

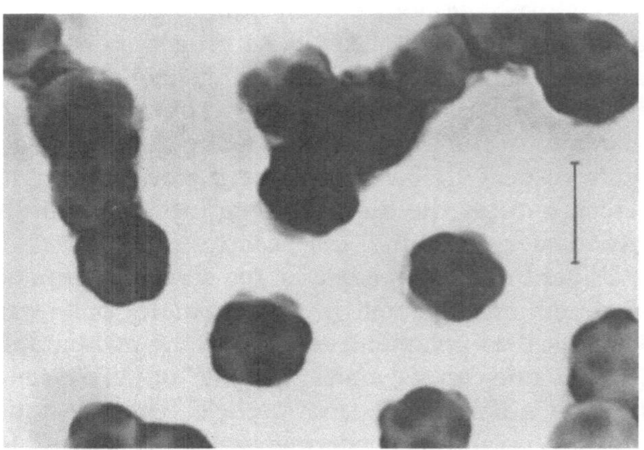

Fig. 1. Electron micrograph of shell formation of PS on a PVAC core latex particle: PVAC: PS = 80:20; the bar indicates 250 nm

Table 1. Average diameter of PS (a) and PMMA (b) islands on PVAC cores

PS fraction %	Diameter nm	PMMA fraction %	Diameter nm
8	57	12	21
13	70	21	28
18	80	23	35
21	90	31	47
31	100—120		
44	115—140		

of the experimental conditions (prevention of new particle generation). The geometry of the core-shell particles is thus determined by the interfacial terms

$$(\Delta G_{INT})_{ik} = O_{ik} \cdot \sigma_{ik} , \quad (3)$$

where O_{ik} and σ_{ik} are the interfacial area and the corresponding interfacial tension between phase i and k, respectively. According to Fig. 2, the interfacial energy can be written as

$$\sum_{i<k} (\Delta G_{INT})_{ik} = O_{I,w}\sigma_{I,w} + O_{II,w}\sigma_{II,w} + O_{I,II}\sigma_{I,II} . \quad (4)$$

Inserting $O_{II,w} \approx O_{I,II}$ and $O_{I,w} + O_{II,w} \approx O_\tau$ into (4) leads to

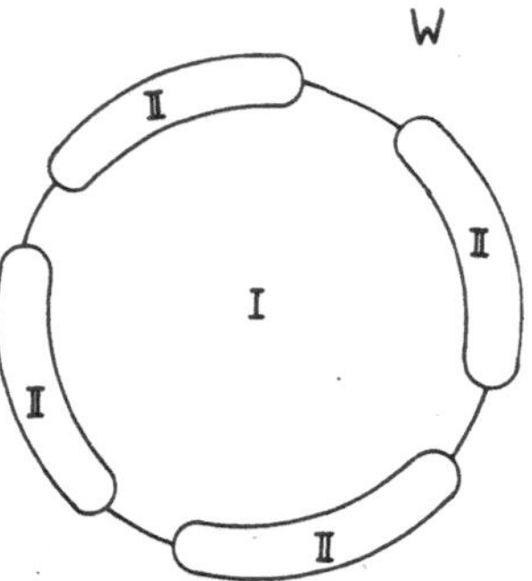

Fig. 2. Scheme of spreading of polymer II on polymer I core (W — water)

$$\sum_{i<k} (\Delta G_{INT})_{ik} = \sigma_{I,w}O_\tau + (\sigma_{II,w} + \sigma_{I,II} - \sigma_{I,w})O_{II,w} , \quad (5)$$

where O_τ is the particle surface. The first term on the righthand side of Eq. (5) is given by the seed latex. Hence, the relation between the interfacial tensions of the components in parentheses of the second term determines the change of interfacial energy. The energy decreases if $(\sigma_{II,w} + \sigma_{I,II}) < \sigma_{I,w}$. In this case (or when $\sigma_{II,w} < \sigma_{I,w}$ holds), the increase of energy per particle is smallest if $O_{II,w}$ is largest. In cases where the core-water interfacial tension exceeds the shell-water interfacial tension the shell polymer should spread perfectly on the core polymer. In the opposite case, the shell polymer will form a very small-sized area against water. This interpretation is consistent with the electron microscope findings, if we compare the monomer-water interfacial tensions as a measure of the corresponding polymer-water interfacial tensions:

$$\sigma(\text{styrene-water}) = 35.0 \text{ mN/m}$$

$$\sigma(\text{MMA — water}) = 18.8 \text{ mN/m}$$

$$\sigma(\text{VAC — water}) = 19.5 \text{ mN/m} . \quad (8)$$

According to these values, PVAC-PS core-shell combination leads to islands, whereas PS-PVAC results in an ideal core-shell structure even with low shell polymer mass fraction.

ii) Beside this spreading effect, the absorption of monomer II and the adsorption of oligomers (and oligoradicals) formed in the aqueous phase is different in separate regions of the core-shell particle.

It is known from the literature [9, 10], that the amount of adsorbed anionic surfactants (having comparable chain length as the oligomers or oligoradicals in emulsion polymerization) on non-polar hydrophobic particle surfaces (e.g., PS) is three to 10 times larger than on more polar, hydrophilic surfaces (e.g., PVAC, PMMA, PMA). Hence, the PS phase is the favored locus of adsorption of the oligomeric radicals of the second polymerization stage in the cases of PVAC-PS and PS-PVAC. This leads to an improved growth of the PS islands in the first case, and an increasing number of small-sized "point" islands in the latter case (which results in more homogeneous spherical shell formation).

The selective absorption of monomer becomes also important if the solubility of monomer II is different in polymers I and II. The Flory-Huggins interaction parameters at $T = 333$ K are, e.g., χ (St — PVAC) = 0.42 and χ (St — PS) = 0.26 [11]. Hence, styrene is a better solvent for PS than for PVAC. Therefore, the monomer concentration within the PS islands should exceed the monomer concentration in the PVAC particle. This also favors the growth of the islands.

IV. Verification of core-shell structure

The polymer I-polymer II combinations used in this section were polymerized according to the discussed procedure for producing perfect core-shell structure.

One of the main problems in investigating core-shell latex particles is the detection of this type of internal structure. The direct observation by electron microscopy does not permit any conclusive evidence for the formation of shells around cores. Ultra-thin cross-sections and selective staining of the samples did not give a better insight into the particle structure by transmission electron microscopy [2].

However, the structure of such particles can be obtained by scattering investigations: small-angle x-ray and neutron scattering. The scattering intensity observed $I(q)$ is directly determined by the square of the scattering amplitude $\Phi(q)$ of a sphere or of the core-shell system because of the low polymer concentration and the high monodispersity of the samples ($\overline{R_w}/\overline{R_N}$ = 1.006 and 1.010 of PS core and PS-d-PMMA, respectively; where $\overline{R_w}$ and $\overline{R_N}$ are the weight and the number average particle radius, respectively). The scattering amplitude Φ_s of N independently scattering solid spheres of scattering density ρ distributed in a dispersion medium of scattering density ρ_u is given by (Fig. 3c):

$$\Phi_s(\Delta\rho, R, q) = 4\pi N \Delta\rho \frac{\sin(qR) - qR \cdot \cos(qR)}{q^3}, \quad (6)$$

where R, q, and $\Delta\rho = \rho_u - \rho$ are the particle radius, the wave vector, and the contrast, respectively. For the general case of a core-shell system (Fig. 3a), the scattering amplitude $\Phi_{cs}(q)$ can be derived from Eq. (6) by superposition:

$$\Phi_{cs}(q) = \Phi_s(\Delta\rho_2, R_2, q) - \Phi_s(\Delta\rho_2 - \Delta\rho_1, R_1, q), \quad (7)$$

with R_2 being the outer shell radius, R_1 the core radius, $\Delta\rho_2$ the contrast of the shell polymer with respect to the dispersion medium, and $\Delta\rho_1$ the contrast of the core polymer with respect to the dispersion medium. For $\Delta\rho_1 = 0$, one yields the known scattering amplitude for a system of hollow spheres [12].

The scattering contrast in SAXS experiments is determined by the corresponding electron densities. Because of the small differences between electron densities of the core polymer, shell polymer and water, this method has only limited application in some special polymer combinations of core-shell systems. Details of the experiments and results have been discussed elsewhere [5].

In SANS experiments, the scattering density ρ can be manipulated by selective deuteration of the components. In particular, the core scattering was matched by using a mixture of H_2O and D_2O in the case where completely deuterated MMA (d-MMA) was used in the second stage, while undeuterated styrene was polymerized as the core. This leads to hollow sphere model system for scattering investigations (Fig. 3b).

In Fig. 4 the SANS profiles of a PS core (a) and a PS-d-PMMA core-shell latex (b) thus obtained are plotted. The radius of the PS core particles is R = 215 nm by SANS, resp. R = 219 nm by light scattering. Using R from SANS as R_1, a shell thickness of 40 nm was derived by fitting the experimental data into Eq. (7) (it was impossible to fit the experimental data with the sattering function of a sphere (Eq. (6)). From the chemical composition of the latex, a shell thickness of about 50 nm can be estimated. The difference may arise from the fact that in Eq. (7) as stepwise density profile is assumed, whereas in reality the concentrations of the core

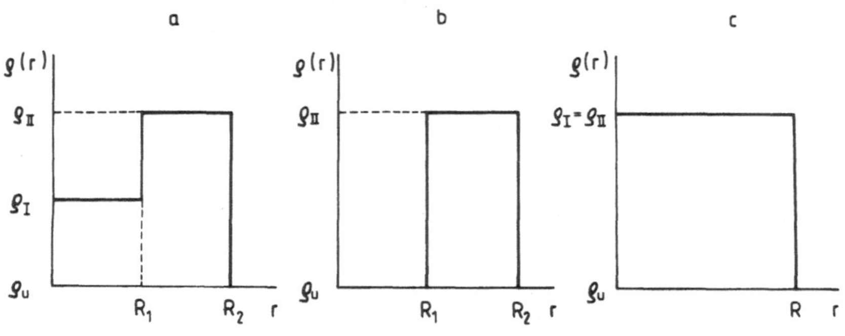

Fig. 3. Radial distribution of scattering density ρ for core-shell system (a), hollow shell system (b), and sphere (c)

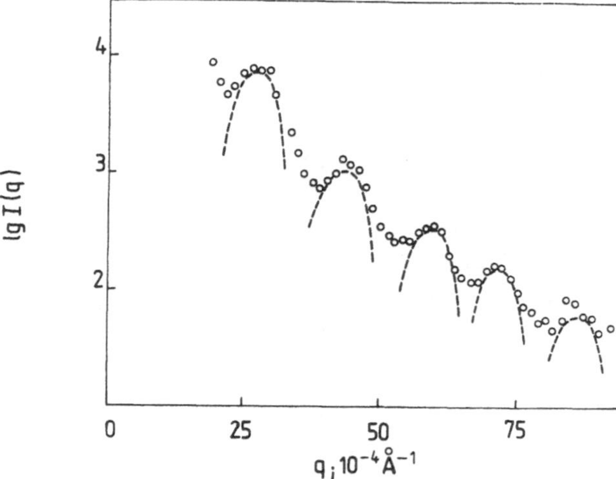

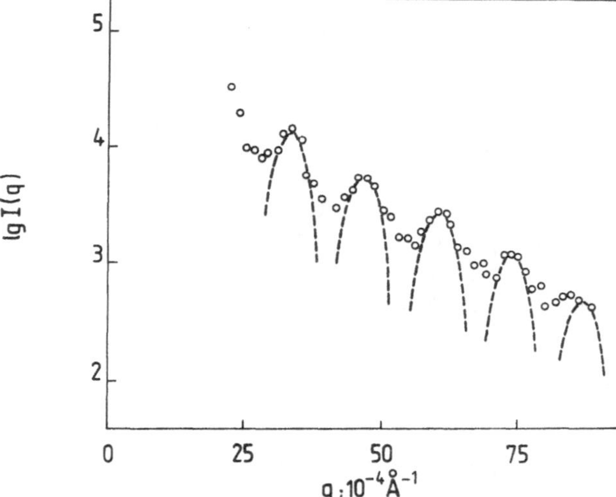

Fig. 4. Neutron scattering intensity as a function of wave vector (desmeared and corrected values): a) Monodisperse PS latex. o — experimental; ——— calculated for sphere of radius 215 nm; b) (PS-core)-(d-PMMA shell) latex. o — experimental; ——— calculated for hollow shell of 40-nm shell thickness

and of the shell polymer should monitonically change inside an interfacial layer between both polymers.

It is obvious from the small angle scattering investigations that the polymer particles produced under the experimental conditions discussed in the previous section exhibit a distinct core-shell structure. This indicates that the surface layer of polymer particles is the main reaction locus during two-stage emulsion polymerization.

V. Core-shell interfacial layer

Va. Polymer-polymer systems

The small interfacial layer between core and shell material (as indicated by scattering experiments) is important for understanding the macroscopic properties of polymer blends, especially deviations from the sum of the ingredients' properties. The direct determination of interfacial layer thickness from detailed extended neutron scattering investigations will be published in a subsequent paper. In the present paper, we discuss the estimation of interdiffusion layer thickness from i) differential scanning calorimetry (DSC), and ii) infrared spectroscopy (IR).

i) The glass transition temperatures T_G of the individual polymers of a two-stage PVAC-PMMA core-shell system did not show any correlation with structural particularities of the particles. The variation of T_G in dependence on polymer composition is in the range of ± 3 K. The separate appearance of the components' glass transition (Fig. 5) indicates that both polymers exist in separate phases [13, 14] as the scattering investigations have shown. The width of the glass transition region is independent of chemical composition. However, a characteristic

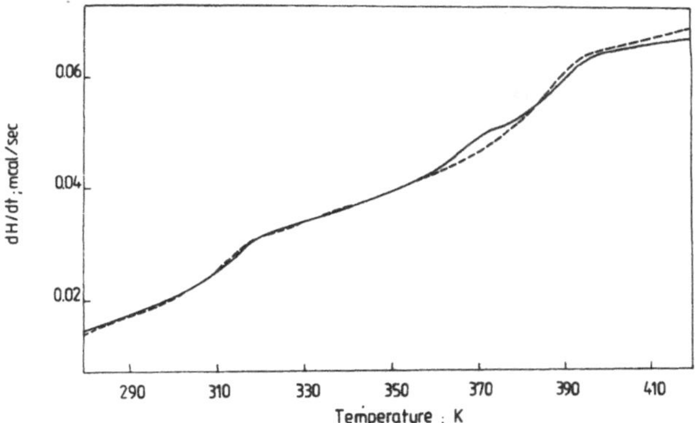

Fig. 5. DSC recording during heating of a PVAC-PMMA core-shell sample with PMMA weight fraction of 0.72. ---- original; —— annealed for 2 h at 345 K

behavior of the increase of specific heat capacity ΔC_p at the glass transition temperature of both polymers was found. Experimental ΔC_p values are lower than theoretical values calculated from the net chemical composition of the latex. This can be understood if we assume that the interdiffusion layer between core and shell polymer forms a separate third part of the system. The chemical composition of this layer can be calculated from the experimentally "missing" amounts of the individual ΔC_ps'. The average weight ratio of both polymers forming this component is about $W(\text{PVAC}):W(\text{PMMA}) \approx 1:3$.

The existence of this third component can be verified by proving relaxation processes in the DSC thermograms between the glass transitions of both polymers. Therefore, the samples were annealed at 345 K for 2 h (see Fig. 5). A weak relaxation process between both glass transitions is clearly observable. This relaxation process is responsible for the missing amounts of ΔC_p because it is "smeared over" between both transitions in the usual DSC thermograms of the un-annealed samples. Possibly, this additional relaxation process is caused by graft products arising from the two-step emulsion polymerization process.

Obviously, the proportion of PVAC chains located in the interfacial layer in the vicinity of the hard glassy PMMA (T_G(PMMA) = 386 K) cannot contribute to the PVAC glass transition (T_G(PVAC) = 314 K), because of their hindered molecular mobility. However, they contribute little by little to the relaxation process between both transitions, and, due to this they soften the PMMA chains in the interdiffusion layer even at temperatures below the PMMA glass transition. Hence, both ΔC_ps of the pure core and shell component are smaller than expected from chemical composition.

Using the excess parts of ΔC_p of both components and considering the geometrical shape of the polymers, one can calculate the volume fraction of the interfacial layer. The corresponding layer thickness between core and shell material is in the order of $2 \text{ nm} < d < 7 \text{ nm}$. For a detailed discussion see [15].

ii) It is known that vibrational spectroscopy is a sensitive method of detecting the structural peculiarities of polymers [16]. Therefore, we carried out IR spectroscopical investigations on PVAC-PS core-shell polymers in order to derive the interfacial layer thickness. The IR spectra of the samples indicate that out-of-plane vibrations of the substituent groups are more strongly influenced by polymer-polymer interactions (i.e., by the nature of the neighbor) than are in-plane or backbone vibrations. Therefore, the intensity of the out-of-plane styrene ring vibration of PS at 700 cm^{-1} was investigated as a function of the PS concentration. The C—C-stretching vibration of the ring at 1607 cm^{-1} was chosen as an internal standard. The ratio of absorbances A (700 cm^{-1})/A (1607 cm^{-1}) as a function of composition is polotted in Fig. 6. This

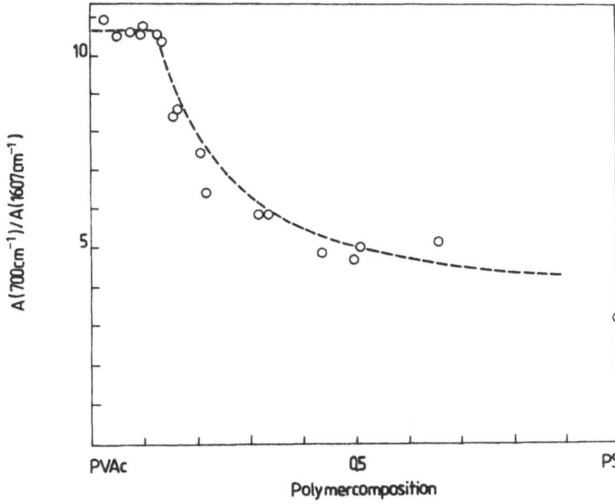

Fig. 6. The ratio A (700 cm^{-1})/A (1607 cm^{-1}) for PVAC-PS core-shell polymer as a function of polymer weight fraction (see text). ○ — experimental; ---- calculated

ratio is constant at 10.5 for small contents of PS from zero up to about 13%, then decreases with increasing PS percentage and approximates to the value which was found for pure bulk PS (3.2). Comparing this with electron microscope observations, it is clear that the composition at which the absorbance ratio starts to decrease is nearly identical to the composition at which the PS islands appear. Therefore, we identify the absorbance ratio 10.5 at very low PS volume fractions with interfacially polymerized PS (i.e., PS polymerized at the PVAC particle surface). During continued polymerization, the interfacially polymerized PS remains nearly constant, whereas the bulk PS portion increases (i.e., PS polymerized in the PS islands).

Assuming the additivity of the absorbances of the interfacial and of the bulk PS characterized by the aforementioned ratios, it is possible to calculate the A (700 cm^{-1})/A (1607 cm^{-1}) ratios as a function of the polymer weight fraction according to Lambert-Beer's law:

$$\frac{A\ (700\ cm^{-1})}{A\ (1607\ cm^{-1})} = \frac{\varepsilon_1 \cdot C_1 + \varepsilon_2 C_2}{\varepsilon_3} \cdot \frac{1}{C_1 + C_2}, \quad (8)$$

where ε_1, ε_2 and ε_3 are the coefficients of absorbance at 700 cm^{-1} of the interfacial PS, at 700 cm^{-1} of pure bulk PS, and at 1607 cm^{-1} as the reference, respectively. C_1 and C_2 are the appropriate weight fractions of interfacial and bulk PS, respectively. The correlation between the calculated and the experimentally obtained absorbance ratios in Fig. 6 is remarkable. Using the value of C_1 derived from the experiments, an interfacial layer thickness of about 5 to 7 nm can be estimated [17]. This value is in the same order of magnitude as the results from DSC investigations.

Vb. Inorganic filler-polymer systems

In general, it is possible to replace the polymeric cores by inorganic filling material [2, 7, 18]. Previously, it was shown that highly-dispersed quartz powder filler particles ("Suprasil" with an average particle diameter $2R$ = 26 nm) can be coated with PVAC by emulsion polymerization [18].

The polymer which originates close to the filler surface is bound to this surface. Hence, the physical properties of such amorphous polymers are different from those of the bulk polymer. Because of the binding to the rigid surface, the molecular mobility of the attached polymer chains is hindered and their molecular conformation is changed. Additionally, the structure of the polymer in the vicinity of the surface should be altered from an amorphous to a higher ordered state as a consequence. We applied DSC and NMR measurements in order to determine the thickness of this modified interfacial layer on the surface of the filler particles.

In general, the width of the glass transition region of the polymer in the silica-PVAC system determined by DSC is not influenced by the amount of filler. The influence of filler concentration on the glass transition temperature and on the increase of specific heat capacity at T_G is shown in Fig. 7. The increase of specific heat capacity at T_G deviates considerably from values calculated from the chemical composition. According to the discussion above, this difference is discussed in terms of an interfacial layer. The portion of the polymer chains bound at the immobile particle surface cannot contribute to the glass transition because its molecular mobility is hindered. This portion lowers the increase of C_p. The decrease is proportional to

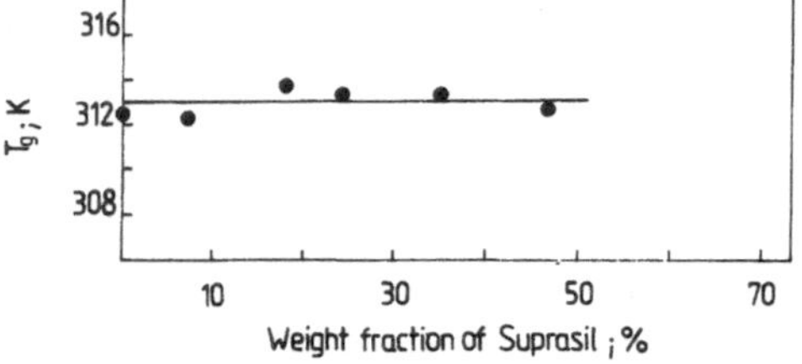

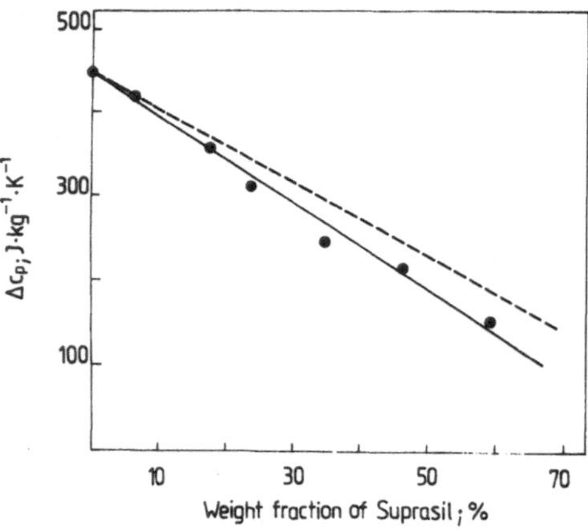

Fig. 7. a) Variation of the glass transition temperature T_G and b) the increase of specific heat capacity ΔC_p of PVAC with the Suprasil weight fraction. ● — experimental; ---- calculated from chemical composition

the filler volume fraction. Using the known filler particle diameter and the $\Delta C_p(0)$ values of the pure polymeric component, the interfacial layer thickness d can be calculated from DSC data

$$\frac{d}{R} = \left[\left(1 - \frac{\Delta C_p}{\Delta C_p(0)(1 - C_f)}\right)\left(\frac{1 - C_f}{C_f}\right) \cdot \frac{\rho_f}{\rho_p} + 1\right]^{1/3} + 1\,, \qquad (9)$$

where C_f is the weight fraction of filler, and ρ_f and ρ_p are the densities of filler particles and polymer, respectively. The resulting interfacial layer thickness is 1.4 ± 0.9 nm, which is in agreement with NMR data [7].

The model of adsorbed polymer chains with hindered molecular mobility is supported by the following experimental findings. We polymerized samples with constant weight fractions of PVAC (81.9 — 78.2%) and silica (18.8 — 17.8) in the presence of different concentrations of an anionic emulsifier (Semisuxol). The corresponding T_Gs and ΔC_ps are listed in Table 2. The glass transition temperatures decrease with increasing emulsifier concentration due to the well-known softening effect of emulsifier. Additionally, the emulsifier influences the relaxation processes in the interfacial layer since the ΔC_p-deviation from the theoretical value vanishes with increasing emulsifier concentration. Hence, all polymer (matrix plus interfacial polymer) contributes to the glass transition at higher emulsifier concentration. This result shows that the polymer is physically adsorbed at the particle surface. Additionally, some of the polymer chains at the surface may be displaced by emulsifier molecules.

Table 2. Glass transition temperature T_G and increase of specific heat capacity ΔC_p of silica-PVAC samples in dependence on emulsifier concentration; (C_{PVAC} = concentration of PVAC)

Emulsifier %	$\Delta C_p/\Delta C_p(0) \cdot C_{PVAC}$	T_G K
0	0.90 ± 0.02	313
0.20	0.91 ± 0.02	314
1.45	0.93 ± 0.02	312
2.10	0.97 ± 0.02	310
4.05	1.01 ± 0.02	302

Acknowledgements

The authors wish to thank their colleagues Dr. D. Beyer (SAXS), Dr. J. Lange (IR), and Dr. G. Simon (NMR) for carrying out experiments and for helpful discussions.

References

1. Hergeth W-D, Bittrich H-J, Eichhorn F, Schlenker S, Schmutzler K, Steinau U-J (1989) Polymer 30:1913
2. Hergeth W-D (1990) B-Thesis, TH Leuna-Merseburg
3. Tanneberger H, Bittrich H-J, Steinau U-J, Hergeth W-D (1987) Wiss Z TH Leuna-Merseburg 29:721
4. Bittrich H-J (1990) B-Thesis, TH Leuna-Merseburg
5. Beyer D, Lebek W, Hergeth W-D, Schmutzler K (1990) Colloid Polym Sci 268:744
6. Schmutzler K (1982) Acta Polymerica 33:455
7. Hergeth W-D, Steinau U-J, Bittrich H-J, Simon G, Schmutzler K (1989) Polymer 30:254
8. Hörnig K (1987) Thesis, TH Leuna-Merseburg
9. Vijayendran BR, Bone T, Gajria C (1981) In: El-Aasser MS, Vanderhoff JW (eds) (1981) Emulsion Polymerization of Vinyl Acetate, Applied Science Publ, London and New York, p 253
10. Vijayendran BR (1979) J Appl Polym Sci 23:733
11. Vanzo F, Marchessault RH, Stannett V (1965) J Colloid Sci 20:62
12. Pecora R, Aragon SR (1974) Chem Phys Lipids 13:1
13. O'Connor KM, Tsaur S-L (1987) J Appl Polym Sci 33:2007
14. El-Aasser MS, Makgawinata T, Misra S, Vanderhoff JW, Pichot C, Llauro MF (1981) In ref 9, p 215
15. Hergeth W-D, Steinau U-J, Bittrich H-J, Tanneberger H (1990) Colloid Polym Sci 268:991
16. Laane J, Durig JR (eds) (1983) Vibrational Spectroscopy and Structure, vol 12, Elsevier, New York
17. Lange J, Hergeth W-D, Wartewig S (1988) Acta Polymerica 39:481
18. Hergeth W-D, Starre P, Schmutzler K, Wartewig S (1988) Polymer 29:1323
19. Lipatov YuS (1977) Physical Chemistry of Filled Polymers (russ), Khimija, Moscow
20. Howard GJ, Shanks RA (1981) J Appl Polym Sci 26:399 and J Macromol Sci — Phys B9:67
21. Alfthan E, DeRuvo A, Rigdal M (1979) Intern J Polym Mat 7:163

Received January 3, 1991
accepted February 8, 1991

Authors' address:

Dr. W.-D. Hergeth
Department of Physics
"Carl Schorlemmer" Technical University
of Leuna-Merseburg
O-4200 Merseburg, FRG

Discussion

WENDORFF:

You might be able to determine the thickness of the interfacial layer in core-shell structure from the tail of the x-ray or neutron-scattering curves. Did you try that?

HERGETH:

Recently, we carried out extensive neutron-scattering studies with core-shell particles. Now, we are able to determine the thickness of this interfacial layer from scattering profiles. For instance, in the case of PS-d-PMMA particles the thickness of the layer thus obtained is about 7 nm. This value is in agreement with the data discussed from DSC and IR investigations. The results will be published in detail in a subsequent paper.

KILIAN:

Did you try to verify the core-shell particle structure by electron microscopy?

HERGETH:

Yes, but there are a lot of difficulties because of the small differences between electron densities of the components. The problems in investigating the core-shell structure by electron microscopy can only be partially overcome by the selective staining of the different polymers and the preparation of ultra-thin cross-sections: i) Selective staining often changes either reaction kinetics (e.g., if monomers are used which are marked with heavy metals) or particle structure (e.g., due to swelling). ii) The probability of cutting the "bottom" and the "top" of a particle using a microtome is very low. Hence, it is necessary to apply an additional method to verify conclusively the core-shell particle structure (e.g., SANS, SAXS).

CHUDACEK:

Are there any restrictions concerning the coating of inorganic particles by emulsion polymerization?

HERGETH:

If an inorganic seed material is used, condition (1) can be applied in the same way as discussed in section III. The number of seed particles should exceed the value N_I. In most cases, emulsion polymerization is carried out in the presence of surfactants. Hence, the resulting higher particle number according to condition (1) requires particle dimension $2R \ll 100$ nm [18]. An additonal difficulty may arise from the fact that most of the inorganic fillers are contaminated with salts of other water soluble material. Hence, reaction kinetics and topochemistry may be changed.

Preparation, structure, and some properties of organosilicon thin polymer films obtained by plasma polymerization

A. M. Wróbel and M. Kryszewski

Centre of Molecular and Macromolecular Studies Polish Academy of Sciences, Łódź, Poland

Abstract: The present paper describes advances in the study of preparation and properties of plasma polymerized organosilicon compounds. Plasma polymerization — deposition process is characterized with regard to the fabrication variables showing the condition at which plasma-induced polymerization or atomic mechanism dominate. The temperature dependences of the film deposition rate revealed that the deposition process is generally controlled by adsorption of film-forming species. The discussion is mostly devoted to the dielectric materials, but some remarks are presented on recently obtained ionically conducting plasma-deposited thin films. Structure of the films characterized by various techniques leads to suggestion of the model of organosilicone films. The influence of fabrication variables on structure, morphology, and selected properties are presented.

Key words: Plasma polymerization; deposition; organosilicon film; structure; properties

1. Introduction

Rapid development of VLSI, ULSI (very- and ultra-large scale integrated circuits), and other modern technologies have created a great need for fabrication of new thin-film materials. This fact has stimulated marked progress in chemical vapor deposition (CVD) processes. Of these techniques, plasma-enhanced chemical vapor deposition (PECVD), due to its many beneficial aspects, attracts particular attention. In the light of bicyclic step-growth mechanism of plasma polymerization postulated by Yasuda [1], PECVD may also be recognized, to some extent, as plasma polymerization. This results from the fact that in plasma polymerization, polymerization and deposition are inseparable components of the polymerization-deposition mechanism [1]. Therefore, the PECVD deposits in terms of their chemical structure may be classified as "plasma polymers". This term may be used in reference to organic, as well as to inorganic materials. On the other hand, taking into account the electronic structure of PECVD films, they may be considered as covalent glasses [2].

An outstanding virtue of PECVD or of simple plasma polymerization is low processing temperature, which usually does not exceed 300 °C (compared with 700 °C for conventional CVD). Moreover, PECVD, in contrast to thermal- and photo-CVD, offers much higher deposition rates and the broad possibility for controlling the structure and properties of the deposit [3].

Plasma-deposited films, owing to their unique properties, have attained considerable importance as new materials for fabricating various thin film devices. Thin-film semiconductors are already used commercially in photovoltaic cells [4, 5], in electrophotography [6] in field-effect transistors [7], and large-area displays [7]. Excellent passivation and dielectric materials such as silicon nitride, silicon oxide, and carbide are produced from silane with various admixtures [3, 8]. A relatively unexplored material which has become increasing important is amorphous carbon a-C:H, deposited from various hydrocarbons. It shows many excellent properties required for electronic components [9]. Conducting films in demand for VLSI and other technologies can be prepared from volatile organometallic compounds [10]. Important progress is observed in plasma-enhanced deposition of metallic and metal oxide films from suitable organometallic precursors mixed with H_2 or O_2 [10—16]. By using mixtures of oxygen with various organometallic precursors, many oxides and high T_c superconductors have been deposited and characterized [17—20].

Of particular interst, since the beginning of plasma chemistry, is the fabrication of thin polymer films from organosilicon source compounds, because: 1) the large chemical family of organosilicons includes abundant group of compounds which are sufficiently volatile near ambient temperature, thus used easily in plasma processing; 2) organosilicons, contrary to many organometallics, tend to be relatively nontoxic and they are generally low- or nonflammable, relatively cheap, and available from commercial sources; 3) conventional organosilicon polymers and elastomers play an important role in macromolecular chemistry, therefore, attention has been directed to synthesis of plasma-polymerized counterparts; and 4) as modern semiconductor technology is largely based on the use of silicon, the natural chemical affinity between pure, monocrystalline silicon (c-Si), and organosilicon plasma polymers has motivated device-oriented research.

The above-mentioned remarks do not suggest that plasma polymerization is only attractive for preparation of dielectric, semiconducting or metallic thin films. This technique has several advantages for preparation of thin, ionically conducting films, hence, we discuss briefly the preparation conditions of new, plasma-deposited, ionically conducting organosilicon films.

The present paper describes recent advances in the study of preparation and properties characteristic of plasma-deposited organosilicon thin films, excluding their electrical properties.

2. General characteristics of plasma polymerization-deposition process

The general features of plasma polymerization are known, but some of them need to be recalled.

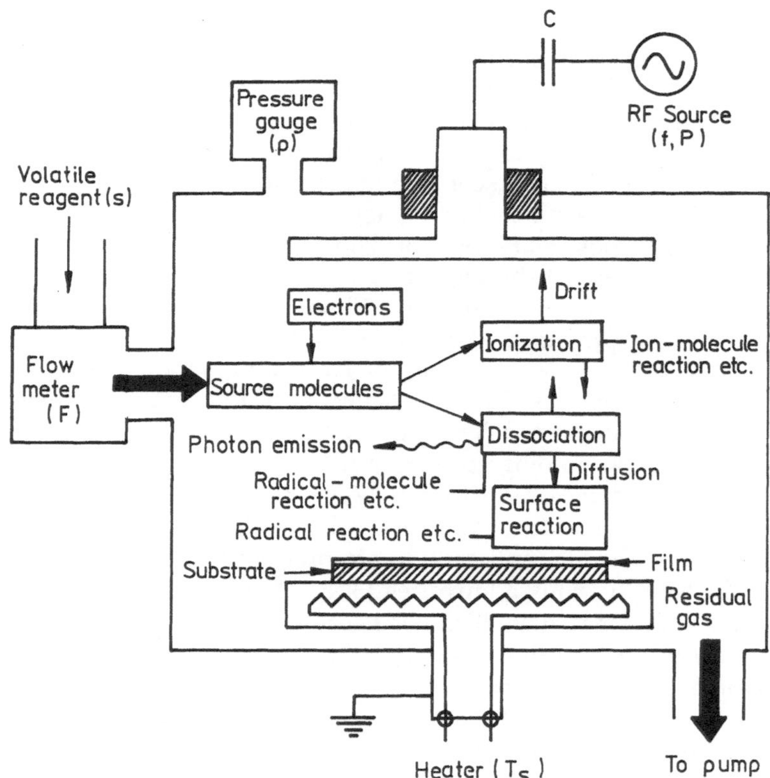

Fig. 1. Scheme of a typical electrode reactor for plasma polymerization and symbolically indicated principal plasma processes

In this process electron-impact dissociation, excitation, and ionization of the gaseous precursors are the primary steps leading to film deposition. The most important phenomena are illustrated in Fig. 1, which presents a scheme of a capacitively coupled rf reactor with internal electrodes. Neutral fragments (radicals) produced in the gas phase diffuse toward the substrae, whereas ionic species drift toward the electrodes under the influence of the applied electric field. Electronically or vibrationally excited neutral species emit light of wavelengths ranging from vacuum UV to IR. A number of secondary processes such as neutral-molecule, ion-molecule, and photon-molecule reactions take place in the gas phase. Finally, heterogeneous chemical reactions of the reactive plasma species impinging onto the surface result in the formation of the deposit.

At a typical operating pressure in rf plasma system $p = 0.1$ torr (13.3 Pa), electron and ion densities are, at most, 10^{11} cm^{-3}, while the number density of molecules is about 10^{15} cm^{-3}. Unlike the ions and molecules which remain essentially at ambient temperature (T_i — 300 K), the average electron energy exceeds 1 eV (or, equivalently, $T_e \sim 10^4$ K). As the threshold energy to create neutral fragments by electron impact is much lower than that needed for ionization, the predominant species impinging onto the substrate surface are inferred to be radicals rather than ions. Therefore, the film-formation process might be controlled either by generation rate of radicals or by the surface radical reactions. Ionic species are generally considered to be less significant in plasma polymerization than are radicals. We have shown that in the case of organosilicon monomers, they also contribute to the crosslinking process in film formation [10, 21], depending on fabrication variables.

3. Effect of fabrication variables and nature of plasma polymerization-deposition process

Plasma polymerization process can be controlled by a number of operational parameters such as: power input P, frequency of electrical energy source f, total pressure p (and reactant partial pressures, in the case of gas mixtures), gas flow rate(s) F, pumping speed, substrate temperature T_s; and such reactor design elements as overall geometry, electrode spacing, electrode material, and whether the substrate is mounted on the powered (upper, in Fig. 1) or grounded (lower) electrode. The reason for this latter distinction is related to self-bias negative potential, which can develop on an electrode due to formation of a plasma sheet near the electrode surface [22]. Substrate can also be biased by using a separate power source. In this case, controlled substrate bias potential V_s constitutes another powerful fabrication variable. All these variables mutually interacting determine the rate of film deposition, as well as properties of the deposit.

Significant progress in precise evaluation of plasma polymerization-deposition conditions has been made by Yasuda and Hirotsu [23], who introduced a composite parameter P/FM (M being the molecular weight of the monomer) which expresses the plasma energy imparted to the system per unit mass of the monomer. The parameter permits to evaluate various types of plasma-polymerization-deposition process [24]. In our recent work, we have used it to determine if the polymerization and deposition rate depends on structural features of the monomer, or if the process of polymerization proceeds via an atomic mechanism.

Plasma polymerization-deposition kinetics is also strongly affected by substrate temperature (it influences the surface mobility of the adsorbed polymerizable species and their sticking probability).

Our reinvestigations of the temperature (T_s) dependence of the deposition rate R_d evaluated for plasma-polymerized (PP) tetramethylsilane (PP-TMS) hexamethylsiloxane (PP-HMDSO) and hexamethyldisilazane PP-HMDSN) [25—26] are present in Fig. 2. The deposition rate is seen to decrease with increasing T_s value. It proves that plasma polymerization-deposition is an adsorption-controlled process. The apparent heats of adsorption calculated from the slopes of plots in Fig. 2 are equal to 9.9; 4.8 and 5.3 kJ/mol for TMS, HMDSO and HMDSN respectively (for T_s in the range 25—250 °C). The role of adsorption has also been confirmed by the good agreement found between the experimental pressure dependence of the R_d and adsorption equation [27]. The separate high-temperature region (200 °C $< T_s \leqslant$ 500 °C) in the plots of PP-HMDSO and PP-HMDSN (Fig. 2) is attributed to thermal decomposition. Very low heats evaluated for this region indicate that the deposition of polymer film in this regime of T_s can be a diffusion-controlled process.

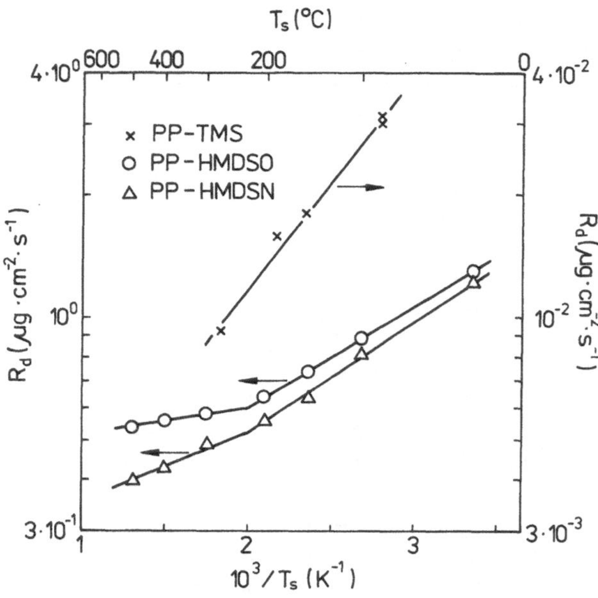

Fig. 2. Arrhenius plots of the relationship between the deposition rate R_d and substrate temperature evaluated for plasma-polymerized tetramethylsilane (PP-TMS) [25], hexamethyldisiloxane (PP-HDMSO) [26], and hexamethyldisilazane (PP-HMDSN) [26]

4. Structure of plasma polymer films

Due to a very small quantity of the plasma polymers produced in usual laboratory reaction (milligrams) and poor solubility of this material in organic solvents (high degree of crosslinking) its structural examination need specific analytical tools.

The structure of plasma polymer films deposited from five main groups of monomers: alkyl- and phenylsilanes, alkoxysilanes, alkylsiloxanes, alkylaminosilanes, and alkylsilazanes have been characterized by IR-spectroscopy, elemental analysis (EA), x-ray photoelectron spectroscopy (XPS), Auger electron spectroscopy (AES), combined pyrolysis/gas chromatography/mass spectrometry (P/GC/MS), solid-state nuclear magnetic resonance (NMR) spectroscopy, electron spin resonance (ESR) spectroscopy, differential scanning calorimetry (DSC), and electron microscopies (TEM and SEM). The major achievements in the structural examinations obtained by using the mentioned techniques will be described in the following section.

4.1 Structural examinations and structure model

The IR examinations of PP-films produced from various groups of organosilicon monomers [21, 22, 25, 26, 28—30] revealed incorporation of the monomer units: $Si(CH_3)_x$, Si—O-Si (in PP-siloxanes) and Si—NH-Si (in PP-silazanes) to the polymer film, as well as contribution of newly formed units SiH_x, Si—CH_2—Si and Si—CH_2—CH_2—Si.

The EA data [31, 32, 33] show two general tendencies in the composition of organosilicon plasma polymers: 1) they are deficient in carbon and hydrogen with respect to the monomer; 2) they contain oxygen, even when this element is absent from the monomers. In the case of silazane monomers, a nitrogen deficiency is noted in the polymers. Similar tendency as in bulk composition of the polymers, determined by EA, was observed in their surface compositions estimated by XPS from the intensities of Si_{2p} or Si_{2s}, C_{1s}, O_{1s}, and N_{1s} peaks [34—37].

Compositional depth profiling of PP-organosilicon films performed by AES technique coupled with sputtering by an Ar ion beam revealed their good compositional uniformity with depth [38].

As proved by P/GC/MS data obtained for low-temperature regime (300°C) of pyrolysis, PP-organosilicons contain low-molecular-weight products trapped in the film (but not bonded chemically to the polymer network) which are mostly not higher than dimer [21, 30, 39, 40]. Contribution of these products, being up to 16—17 wt.%, strongly depends on the plasma parameters, and drastically drops with plasma power [41]. Evolution of ethylene, among other hydrocarbons (CH_4, C_2H_6), observed by P/GC/MS for high-temperature regime (500°C) originated mostly from scision of the Si—C bonds in Si—CH_2—CH_2—Si crosslinks in the polymer [29, 30].

A general structural feature of plasma polymers which suggests the presence of dangling bonds, is relatively high concentration of trapped free radicals. The spin density in PP-organosilicons, evaluated by ESR spectroscopy, ranges from 1.5 to 5×10^{18} spins/cm^3 [42]. High contents of radicals in plasma polymer films is very undesirable due to rapid oxidation following film exposure to the ambient and resulting deterioration of some of its properties. The radical concentration may substantially be reduced, either by deposition of the film in a pulsed plasma [42], or by its annealing in vacuum

[31, 43] or in an inert atmosphere [38], thus enhancing the recombination of trapped radicals by thermal excitation.

Solid-state NMR spectroscopy appears to be a powerful technique for the study of plasma polymer, and has provided quantitative information on the structure of PP-organosilicon [44—46]. The solid-state ^{13}C- and ^{29}Si-NMR data for PP-HMDSO (obtained using magic-angle spinning and cross-polarization techniques) show that this plasma polymer is mostly composed of silylmethyl, silylmethylene, and mono-, di-, tri- and tetrafunctional siloxane units. A marked broadening observed for some signals accounts for various conformations of the branched and crosslinked structure. On the basis of NMR results, the hypothetical model of the structure for PP-HMDSO film, presented in Fig. 3, has been proposed [45]. This structure however, is incomplete as it lacks Si—CH_2—Si crosslinks, the formation of which is considered to be a predominant process in plasma polymerization of methylsilicons [21, 26, 29, 30, 41]. Moreover, GC/MS study of plasma reactivity of silylmethyl groups [28] and solid-state NMR examination of PP-HMSO [46] allow to assume that the concentrations of the disilylmethylene, disilane, and disilylethylene crosslinks in PP-methylsilicons remain in the order | Si—CH_2—Si | > | Si—Si | > | Si—CH_2—CH_2—Si |.

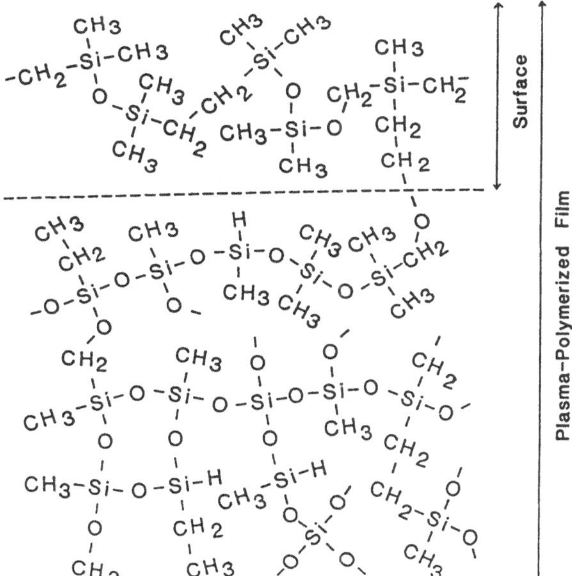

Fig. 3. Model of the structure of PP-Hexamethyldisiloxane film [45]

4.2 Effect of fabrication variables

The structure and composition of plasma polymer films can be selectively modified by a suitable change of certain plasma variables. The most substantial effects are achieved by varying the discharge power, substrate temperature, substrate bias voltage, and concentration of reactive feed gas. The influence of the farbrication variables on the structure of PP-organosilicon film may be exemplified by the effect of discharge power and substrate temperature. Both these parameters appear to have a similar influence upon the structure of PP-film, although the mechanisms involved are different. When P is the variable, the film structure is mainly affected by monomer fragmentation, whereas when T_s is raised, pyrolysis of the growing film and surface mobility of the precursor species on the substrate are the factors influencing the resulting film structure. Increase in P and T_s values leads to intense crosslinking in PP-methylsilicons via Si—CH_2—S links [26, 29].

Typical effects of varying P and T_s are confirmed by the studies of IR spectra of PP-HMDSN films [26, 47]. The intensities of absorption bonds at 2100 (SiH), 1410 (CH_3), 1260 ($SiCH_3$), and 1170 cm^{-1} (NH in SiNHSi) decrease markedly with increasing P and T_s, indicating the elimination of hydrogen and of carbon-containing groups. A broad, intense band which appears at high P and T_s in the range 1100—600 cm^{-1} is characteristic of the strongly crosslinked material with a high content of ionorganic structure. This is consistent with EA results which revealed a significant drop in C and H concentrations with increasing P [32] and T_s [47]. The results presented evidently prove that P and T_s can be used as independent variables, in order to control the organic content in PP-organosilicon films.

Plasma polymerization was recently applied for obtaining ionically conducting plasma polymerized films (see e.g., [49]). The octamethylcyclotetrasiloxane OMCTS was selected as basic monomeric substance. Solid PP-film was obtained in typical plasma polymerization equipment with RF power supply (13.56 MHz). OMCTS was introduced with Ar as transport gas. The thickness of the films without pinholes was 15 μm (SEM).

The specific advantage of polysiloxane is its low polarity. The interactions between ions and lone pairs of electrons from oxygen atoms of the siloxane group cannot cause salt dissociation, and thus cannot generate a significant number of mobile ionic charge carriers. Thin plasma polymerized films obtained from OMCTS were soaked in butanol-PPO solution containing various concentrations of lithium perchlorate at 50 °C. The hybrid polymer film consisting of plasma polymer (75%) PPO (22.5%) lithiumperchlorate (2.5%) after drying under reduced pressure at room temperature to evaporate butanol) exhibits rather high ionic conductivity of the order of 10^{-6} S/cm at 50 °C.

In order to obtain solid plasma polymerized film without plasticizer, we have used the same monomer as Ogumi et al. [49]: Tris(2-metoxy ethoxy)vinylsilane (TMVS). This monomer contains both ether groups and a siloxane group. Thus, one could expect to obtain TMVS films with lithium perchlorate dispersed in the thin layer using multiple plasma polymerization technique for films appropriately treated with lithium perchlorate and heating. We have compared IR spectra of obtained films using the parameter P/FM (previous section). At higher P/FM values the height of peaks decreased while their width increased, except for the region between 1050—1250 cm^{-1}, corresponding to Si—O-Si stretching.

Room temperature conductivities have been greater than 10^{-6} S/cm. Logσ is plotted vs P/FM at different temperatures to show the correlation with glass transition temperature. Usually, under low P/FM condition the conductivity is higher, but our results show much stronger dependence on P/FM than that presented by Ogumi et al. [49]. This means that the crosslinking, for example, —Si—O—Si— bonds formation was much more effective in both stages of ionically conducting film preparation.

4.3 Morphological structure

A general feature of plasma polymer films is the apperance of a two-phase morphological structure under certain plasma conditions (e.g., high discharge power density [27, 50]) comprising spherical or spheroidal particles embedded in a continuous polymer matrix. Whereas the size of the organosilicon powder particles essentially does not exceed 1 μm in diameter [27, 51], their mere presence has a deteriorating effect on the quality of the films in terms of their practical properties. This morphological heterogeneity can be avoided by a suitable selection of the fabrication variables.

The important parameters which can be used to control film morphology are substrate temperature [48] and substrate bias [52]. Effect of T_s on the morphology of PP-HMDSN films on glass substrate [48] is that the film deposited at T_s = 25 °C consists of an agglomerate of spheroidal particles. It was also demonstrated that even a moderate rise in T_s to 200 °C significantly reduces the number of large spheroidal particles and enhances the film smoothness. At higher T_s value (400 °C) the occurrence of particles is almost completely eliminated and the film structure becomes homogeneous. The observed effect results from complex interactions between the plasma and the surface of the growing film. The most important of these interactions is that the sticking coefficient of impinging precursor species decreases and their surface mobility increases with rising T_s. Heat transport into the plasma presumably retards nucleation and growth of powder particles in the gas phase, thus promoting the formation of the homogeneous and more densely packed film structure.

5. Properties of plasma polymer films

Organosilicon plasma polymer films are characterized by many unique thermal, mechanical, optical, electrical, electronic, and electrochemical properties which render them attractive for numerous practical uses. The possibility of wide-ranging structural modification of organosilicon films implies that their properties can also be varied extensively by suitable choice of the fabrication variables. Some selected structural properties of these materials will be discussed below.

5.1 Density profile and effect of plasma UV radiation

The density profile evaluated for PP-organosilicon films revealed heterogeneous film structure. This is exemplified in Fig. 4 [53], which illustrates the local polymer density $\rho(x)$ as a function of depth x in PP-HMDSO film. The density of the topmost, thin (~500 Å) oligomeric layer varies continuously from ρ_0 = 0.76 g/cm^3 (the density of the liquid monomer) to that of the dense, highly crosslinked bulk material. Since the latter must initially have been oligomeric, one can conclude that a relatively slow (of the order of seconds) transformation process of the former to the latter material takes place

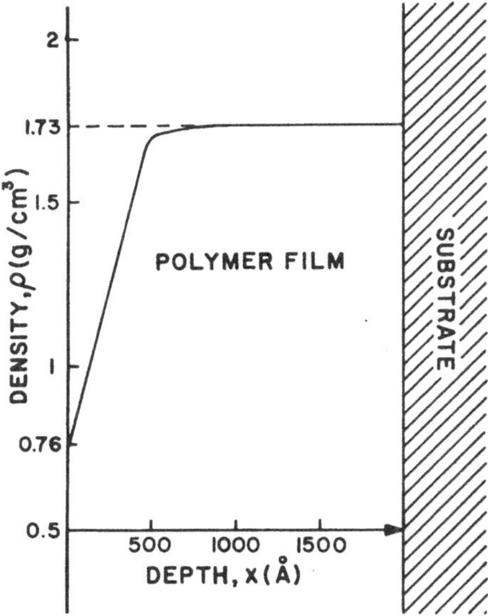

Fig. 4. Profile of the local polymer density $p(x)$ vs depth x in PP-hexamethyldisiloxane film [53]

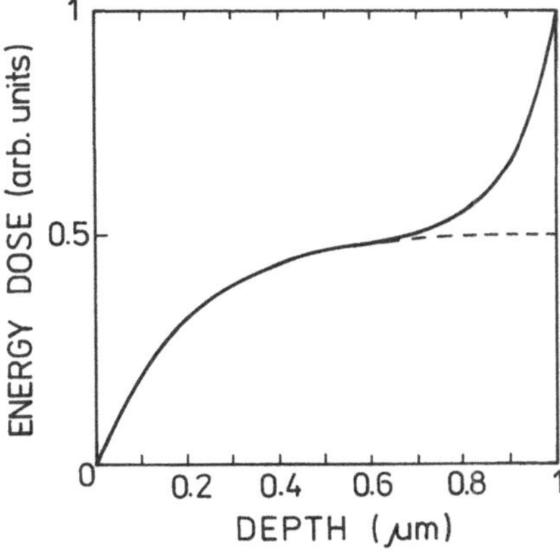

Fig. 5. Energy dose of absorbed plasma UV radiation as a function of the depth in a 1-μm-thick PP-hexamethyldisiloxane film calculated for reflecting (continuous line) and non-reflecting (dashed line) substrates, assuming absorption coefficient $\alpha = 5 \times 10^6$ m^{-1} [54]

in the plasma. This transformation involves a combination of at least two processes: photochemical crosslinking induced by energetic UV photons from the plasma, and outward diffusion of the resulting low molecular weight photolysis products through the material undergoing crosslinking.

The observed density profile is consistent with the calculated profile of the energy dose of plasma UV radiation absorbed in PP-HMDSO film shown in Fig. 5 [54]. As can be noted, the profile for non-reflecting substrate has similar character as that found for the polymer density. This evidently confirms a predominant role of plasma UV radiation in the crosslinking of polymer film. Moreover, the drastic increase in the absorbed dose of UV radiation noted in the polymer layer adjacent to the reflecting substrate (Fig. 5) may substantially influence mechanical properties of the film such as adhesion, elasticity, and internal stress.

5.2 *Wettability and surface free energy*

Measurements of the contact angle θ of water and other liquids of differing polarity revealed that PP-organosilicon films are poorly wettable, low surface free energy materials, generally even more hydrophobic than PP-hydrocarbon films [42, 55]. In some instances, contact angles for PP-HMDSO films deposited at high plasma power densities were found to be in the range $120° < \theta_{H_2O} < 180°$, which implies that they are more hydrophobic than conventional or PP-fluorocarbon polymers [56].

The wettability and surface free energy properties of PP-organosilicon films can be modified by a change in the polymerization conditions. For example, a strong effect on surface polarity is observed when certain gases are admixed to the monomer. This is exemplified in Table 1 [57, 58], which presents the dispersive γ_s^d and polar γ_s^p components of the surface free energy γ_s of the films deposited

Table 1. Effect of admixed feed gas to tetramethylsilane on the surface free energy of the resulting polymer films [57, 58]

Monomer or mixture	Molar ratio	Surface energy (N/m)		
		γ_S^d	γ_S^p	γ_S^*
Me_4Si	—	33.1	1.3	34.4
Me_4Si/O_2	3/1	28.6	1.8	30.4
Me_4Si/O_2	1/1	33.9	5.2	39.1
Me_4Si/O_2	1/3	28.8	23.8	52.6
Me_4Si/N_2	1/2	19.6	19.6	31.8

from tetramethylsilane and its various mixtures with oxygen or nitrogen. These data clearly show that γ_s^p increases with increasing concentration of oxygen or contribution of nitrogen to the reaction mixture. The quite large γ_s^p values observed are due to the formation of O- or N-containing groups in the polymer film [57, 58].

5.3 Mechanical properties

A characteristic feature of PP-films which is attributed to the nature of plasma deposition process is the appearance in these materials of compressive internal stress. Although its origin has not been fully elucidated, it is speculated that reactive plasma species (ions, free radicals, and excited molecules) which continuously impinge onto the growing film, frequently become wedged between existing polymer chain segments, and give rise to the observed compressive stress [59]. In accordance with our recent findings mentioned earlier, a drastic increase in the absorbed dose of plasma UV radiation in the layer adjacent to the substrate (in the case of its reflecting mode, Fig. 5), which implies higher polymer crosslink density in this region, is considered to play a substantial role in creation of the internal stress.

Internal stress value in PP-organosilicon films are typically in the range 0—10 MPa [43, 59, 60], which is an order of magnitude lower than the respective values for PP-hydrocarbon films [43, 59]. Since internal stress often causes buckling and cracking of the film deposited onto the rigid substrate, its value should be reduced by suitable choice of the fabrication or post-treatment conditions, e.g., by annealing at a certain temperature.

The elasticity of PP-organosilicon films can be predicted from the value of mean atomic coordination number z calculated from the continuous random network (crn) structural model proposed for these materials by Gerstenberg et al. [61]. The z value was found to range from 2.2 to 3.2; the former value is typical for elastic polymeric materials (e.g., polyethylene: $z = 2.0$), while the latter corresponds to that of inorganic amorphous solid (e.g., a-Si_3N_4: $z = 3.4$). A structual transition from the elastic to stiff materials is expected at the bond percolation threshold near $z = 2.4$, where all atoms are interconnected by strong covalent bonds. Measurements of Young's modulus E on 1-µm-thick PP-HDSN films confirms this expected transition: E rises sharply from a few GPa at $z = 2.2$ to about 45 GPa at $z = 2.5$ [61]. One of the most important properties of PP-films is their adhesion to the substrate. Deposition rate, film thickness, and substrate surface are of great importance. For instance, the evaluated effect of the polymeric substrate material on the lap-shear adhesive strength of PP-tetramethylsilane films shows that a polyethylene substrate displays the largest adhesive strength (3.73 MPa), followed by poly(methyl methacrylate) (3.14 MPa), and polycarbonate (2.55 MPa), whereas polytetrafluoroethylene has the lowest value (1.08 MPa) [58]. PP-tetramethyldisiloxane films exhibit excellent adhesion to glass, but they adhere poorly to platinum [36]. The reason for good adhesion in the former case is the chemical bonding between the silicon atoms of the plasma polymer and those of the glass via formation of Si—O—Si linkages [36, 62].

5.4 Thermal stability

Owing to their unique structural features, plasma polymers behave differently from conventional polymers during thermal decomposition [30]. Thermogravimetric analysis (TGA) curves of PP-methylsilazanes presented in Fig. 6 reveal three distinct stages in thermal decomposition process. The first stage, at relatively low temperatures up to about 250 °C, is characterized by rapid weight loss due to volatilization of low-molecular-weight products whose presence in the polymer films has been proved by P/GC/MS. In the second stage, which proceeds at temperatures up to about 600 °C, the elimination of organic groups from the polymer (scission of Si—CH_3, Si—$(CH_2)_2$—Si, Si—H, and NH bonds) resulting in the evolution of hydrogen, methane, ethane, and ethylene, takes place. The weight loss observed during the third stage up to about 780 °C, is mostly due to dehydrogenation of Si—CH_2—Si crosslinks remaining in the structure, which results in their conversion to inorganic Si—C bonds [29]. Moreover, the decomposition rate and final weight loss of a particular polymer appear to decrease with increasing number of thermally stable Si—N bonds in the corresponding monomer molecule. The most stable behavior revealed by PP-NN'bis(dimethylsilyl)tetramethylyclodisilzane (curve A, Fig. 6) is attributed to the high contents of thermally resistant N-silyl-substitute cyclodisilazane units in this material [30].

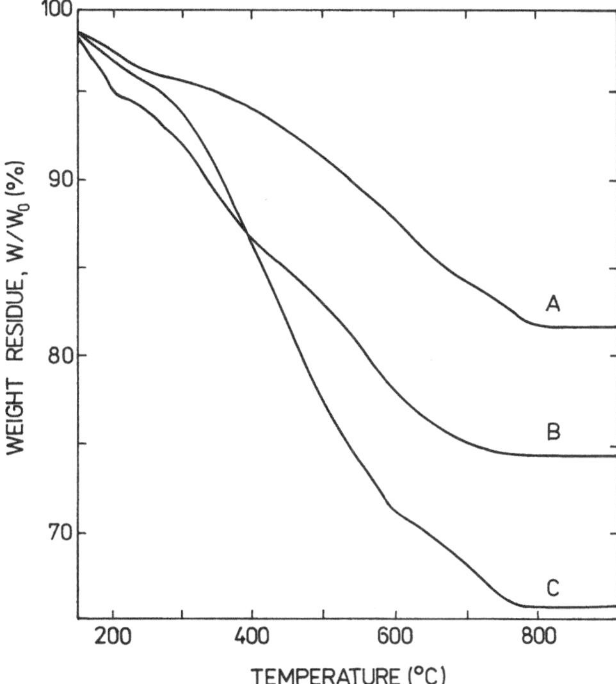

Fig. 6. TGA thermograms of plasma polymers produced from: (A)NN' bis(dimethylsilyl) tetramethylcyclodisilazane, (B) hexamethylcyclotrisilazane, and (C) hexamethyldisilazane, recorded at a heating rate of 8 °C/min in argon

The high thermal stabilities and low degree of decomposition found for PP-organosilicon films enable their conversion into inorganic coatings of superior thermal and chemical resistance by a suitable post-deposition pyrolysis process [63].

6. Conclusions

Recent advances in the study of preparation and properties of plasma polymerized organosilicon films show that, at low power input, molecular polymerization dominates, whereas at high values of this parameter an atomic mechanism is of great importance. The temperature dependencies of the film deposition rate revealed that the deposition process is generally controlled by absorption of film-forming species. The organic content in the film may drastically be reduced by an increase in the substrate temperature or discharge power.

The film characterization by various techniques revealed several important physical properties of the organosilicon films. The density profile confirms a predominant role of plasma UV radiation in the crosslinking of polymer film, however, the internal stresses in PP-organosilicon films are usually smaller than observed for PP-hydrocarbon films. Poor wettability of the PP-organosilicon films can be modified by admixing certain gases to the monomer. The high thermal stability and low degree of decomposition, correlated with a number of Si—N bonds in the monomer, enable conversion of PP-organosilicon films into inorganic coatings of superior thermal and chemical resistance.

References

1. Yasuda H (1985) Plasma Polymerization. Academic Press, Orlando, Florida, Chaps 6 and 8
2. Tyczkowski J (1990) Thin Films of Plasma Polymers (in polish) Polish Scientific-Technical Publishers, Warsaw, Chap 7
3. John P, Jones BL (1987) In: Moss SJ, Ledwith A (eds) The Chemistry of the Semiconductor Industry, Blackie, Glasgow, Chap 6
4. Zweibel K (1986) Chem Eng News, 64:34
5. Kuwano Y (1987) In: Mort J, Jansen F (eds) Plasma Deposited Thin Films. CRC Press, Boca Raton, Florida, Chap 6
6. Mort J, Janson F (1987) In: ibid, Chap 7
7. Matsumura M (1987) In: ibid, Chap 8
8. Adams AC (1987) In: ibid, Chap 5
9. Angus JC, Koidl P, Damitz S (1987) In: ibid, Chap 4
10. Wróbel AM, Wertheimer M (1990) In: d'Agostino R (ed) Plasma Deposition, Treatment, and Etching of Polymers Academic Press, Chap 3
11. Suhr H (1988) Metalloberfläche 42:467
12. Oehr C, Suhr H (1989) Appl Phys A45:151
13. Etspüler A, Suhr H (1989) Appl Phys A48:373
14. Oehr C, Suhr H (1989) Appl Phys A49:691
15. Feuer E, Suhr H (1987) Appl Phys A44:171
16. Suhr H, Etspüler A, Feurer E, Kraus S (1989) Plasma Chem Plasma Process 9:217
17. Holzschuh H, Oehr C, Suhr H, Weber A (1988) Modern Phys Lett 82:1253
18. Koinuma H, Fukuda K, Kogoma M, Okazaki S, Hashimoto T, Kawaski M, Yoshimoto M (1989) Proc IUPAC 9th Int Symp Plasma Chem, Pugnochiuso, Italy, 3:1521
19. Weber A, Suhr H (1989) Modern Phys Lett B13:1001
20. Ebinhara K, Kanazawa S, Ikegami T, Shiga M (1989) Proc IUPAC 9th Int Symp Plasma Chem, Pugnochiuso, Italy, 3:1509
21. Wróbel AM, Kryszewski M, Gazicki M (1983) J Macromol Sci-Chem A20:583
22. Chapman B (1980) Glow Discharge Processes, Wiley-Interscience, New York
23. Yasuda H, Hirotsu T (1978) J Polym Sci Polym Chem Ed 16:743

24. Sharma AK (1986) J Polym Sci A Polym Chem Ed 24:3077
25. Favia P, Colaprico V, d'Agostino R (1989) Proc IUPAC Int Symp Plasma Chem 3:1212
26. Wróbel AM, Klemberg JE, Wertheimer MR, Schreiber HP (1981) J Macromol Sci-Chem A15:197
27. Wróbel AM, Kryszewski M, Gazicki M (1976) Polymer 17:673
28. Wróbel AM, Czeremuszkin G, Szymanowski H, Kowalski J (1990) Plasma Chem Plasma Process 10:277
29. Wróbel AM, Kowalski J, Grębowicz J, Kryszewski M (1982) J Macromol Sci-Chem A17:433
30. Wróbel AM (1987) Plasma Chem Plasma Process 7:429
31. Vasile MJ, Smolinsky G (1972) J Electrochem Soc 119:451
32. Wróbel AM, Kryszewski M, Gazicki M (1976) Polymer 17:678
33. Szeto R, Hess DW (1981) J Appl Phys 52:903
34. Hirotsu T (1979) J Appl Polym Sci 24:1957
35. Sachder KG, Sachder HS (1983) Thin Solid Films 107:245
36. Sharma AK, Yasuda H (1983) Thin Solid Films 110:171
37. Oelhafen P, Cutro JA, Haller I (1984) J Electron Spectrosc 34:105
38. Nguyen VS, Underhill J, Fridman S, Pan P (1985) J Electrochem Soc 132:1925
39. Gazicki M, Wróbel AM, Kryszewski M (1984) J Appl Polym Sci Appl Polym Symp 38:1
40. Wróbel AM (1989) J Macromol Sci-Chem A26:743
41. Gazicki M, Wróbel AM, Kryszewski M (1977) J Appl Polym Sci 21:2013
42. Yasuda H, Hsu T (1977) J Polym Sci, Polym Chem Ed 15:81
43. Marinaka A, Asano Y (1982) J Appl Polym Sci 27:2139
44. Assink RA, Hays AK, Bild RW (1985) J Vac Sci Technol A3:2629
45. Tajima I, Yamamoto M (1985) J Polym Sci, Polym Chem Ed 23:615
46. Tajima I, Yamamoto M (1987) J Polym Sci, A Polym Chem 25:1737
47. Coopes IH, Griesser HJ (1989) J Appl Polym Sci 37:3413
48. Wertheimer MR, Klemberg-Sapieha JE, Schreiber HP (1984) Thin Solid Films 115:109
49. Ogumi Z, Ushimoto Y, Takehara Z (1990) ACS Division of Polymeric Materials: Science and Eng 62:363
50. Kryszewski M, Wróbel AM, in preparation
51. Grębowicz J, Pakuła T, Wróbel AM, Kryszewski M (1980) Thin Solid Films 65:351
52. Martinu L, Klemberg-Sapieha JE, Wertheimer MR (1989) Appl Phys Lett 54:2645
53. Wróbel AM, Wertheimer MR, Dib J, Schreiber HP (1980) J Macromol Sci-Chem A14:321
54. Czeremuszkin G, Wróbel AM, unpublished results
55. Wróbel AM (1983) In: Mittal KL (ed) Physicochemical Aspects of Polymer Surfaces, Plenum Press, New York, Vol 1, pp 198
56. Sacher E, Klemberg-Sapieha J, Schreiber HP, Wertheimer MR (1986) In: Leyden DE (ed) Chemically Modified Surfaces. Gordon and Breach Science Publishers, New York, Vol 1, pp 189
57. Inagaki N, Yagi T, Katsuura K (1982) Eur Polym J 18:621
58. Chen KS, Inagaki N, Katsuura K (1982) J Appl Polym Sci 27:4655
59. Yasuda H, Hirotsu T, Olf HG (1977) J Appl Polym Sci 21:3179
60. Hays AK (1981) Thin Solid Films 84:401
61. Gerstenberg KW, Taube K (1989) Fesenius Z Anal Chem 333:313
62. Yasuda H, Sharma AK, Hale EB, James WJ (1982) J Adhesion 13:269
63. Wróbel AM, Kryszewski M (1979) ACS Symp Ser 108:237

Authors' address:

Prof. Dr. M. Kryszewski
Center of Molecular and Macromolecular Studies
Polish Academy of Sciences
Sienkiewicza 112
90-363 Lódź, Poland

Discussion

STAMM:

You have mentioned that your films are highly crosslinked. Is it, however, possible to control the degree of crosslinking and, in particular, to obtain solvable polymers by plasma deposition?

KRYSZWESKI:

The degree of crosslinking in the plasma polymer films can be controlled most effectively by discharge power and concentration of inert gas in mixture with monomer vapor. These parameters influence fragmentation of the monomer molecules as well as activation of surface of the growing film, thus controlling generation of active sites capable of crosslinking. The films deposited at low discharge power contain large amount of soluble, oligomeric fraction, whereas those produced at high power levels almost exclusively consist of highly crosslinked insoluble polymeric material.

Soluble, linear plasma polymer is known to be produced from tetrafluoroethylene using deposition onto an electrode placed in the field free zone, i.e., inside a Faraday cage made of metallic mesh (e.g., Buzzard D, Soong DS, Bell AT (1982) J Appl Polym Sci 27:3956).

KILIAN:
Could you comment on the qeustion of whether the fractal structure of the deposits is different for a glass obtained by quenching a supercooled melt?

KRYSZEWSKI:
The structure of the plasma polymers differs from that of a glass obtained by quenching of supercooled melt because of high crosslink density. The structure of plasma polymers often resembles that of covalent glasses.

HAVRANEK:
Can you tell us something more about the calculations of the E-modulus of layers?

KRYSZEWSKI:
Young's modulus of plasma polymer was measured by means of a nanometerindenter instrument (Wierenga PE, Franken AJ (1984) J Appl Phys 55:4244). One can also determine Young's modulus by measuring the radius of the curvature of substrate coated with plasma film, that being induced by the internal stress in the plasma polymer (Morinaka A, Asano Y (1982) J Appl Polym Sci 27:2139).

SCHRADER:
Are you able to produce very small metal clusters of a controlled size in the plasma polymers by this technique?

KRYSZEWSKI:
The size of metal clusters is difficult to control in a precise manner. It is dependent on the rate of metal evaporation or the rate of sputtering, as well by the plasma conditions. In particular cases, we can obtain layers with small dispersion of metal grains characteristic of optical absorption in the expected range of UV-VIS spectrum.

STAMM:
When you use a mixed plasma or coevaporation. I could imagine that you can produce "forced blends" of polymers which are incompatible and could not be mixed on a molecular level. Have experiments in this direction already been performed and are those blends stable at room temperature?

KRYSZEWSKI:
In plasma polymerization it is difficult to produce polymer blends. Using two or more monomers (coevaporation), one obtains "copolymer" without specific distribution of inclusions of "homopolymer" dispersed in the matrix of another chemical structure. This is due to the plasma conditions under which all introduced species becomes highly reactive, thus leading to the formation of crosslinked material. This is the reason that plasma polymers differ from the polymer blends.

Deposition, structure, and properties of plasma polymer metal composite films

A. Heilmann and C. Hamann

University of Technology Chemnitz, Department of Physics, Chemnitz, FRG

Abstract: The preparation of plasma polymer thin films containing small metal or semiconductor particles deposited by simultaneous plasma polymerization and metal evaporation is given. Solid-state analytical methods show that metal or semiconductor clusters are encapsulated inside the polymer matrix as microcrystallites or as amorphous materials. The amount and the size distribution of clusters show strong correlation to the electrical and optical properties. These properties can be drastically changed by annealing processes due to coalescence of neighbored cluster configurations. Interesting results, yielded by experiments of suppressed coalescence and of structural changes, are applicable to information storage, e.g., for holographic recording.

Key words: Plasma polymer; polymer films; composite; metal particles

1. Introduction

The aim of plasma polymer metal composite film preparation is the connection of advantages of plasma polymer thin films as adhesive and high resistive media [1] with special properties of thin discontinuous metal films. By encapsulation of metal particles into a polymer matrix, it is possible to get conducting composite films or films with color filter properties [2].

There are different ways to prepare metal-containing plasma polymer films, e.g., by simultaneous metal sputtering and plasma polymerization [3, 4], by simultaneous plasma polymerization and metal evaporation [5, 6], by plasma polymerization of organometallic monomers, or by addition of metallic compounds to the monomer used for plasma polymerization [7, 8]. A great variety of metal-containing plasma polymer films can be achieved by using different monomers and metals. If the film contains a mixture of metal and polymer without chemical bonds between, then the term composite film is used.

In this paper, we present the preparation of composite films by simultaneous plasma polymerization and metal evaporation and the structural, electrical, and optical properties, as well as the change of these properties by thermally induced coalescence processes.

2. Preparation

In a vacuum bell jar (Fig. 1) plasma polymerization and metal evaporation can be carried out simultaneously. Two electrodes (1, 2) with a diameter of 160 mm are used for a.c. or d.c. glow discharge. The metal evaporator (3) was situated under the middle of the lower electrode. The metal was evaporated from tungsten boats or wires. Three teflon bars (4) support electrodes and evaporator. As monomers for the plasma polymerization, benzene or some of its derivatives and hexamethyldisilazan (HMDS) were used. The voltage can be varied between 1000 V and 3000 V, and a typical monomer flow rate of 0.1 Pa l s^{-1} was used. The films were deposited at room temperature. We succeeded in preparation of plasma polymer metal composite films containing Ag, Au, Cu, In, Ge, Co, Sn, Bi or Ho.

The filling factor f is defined by the volume part of the metal in the whole film. The deposition of films with constant or with spatially varying filling factor is possible. Films with nearly constant filling

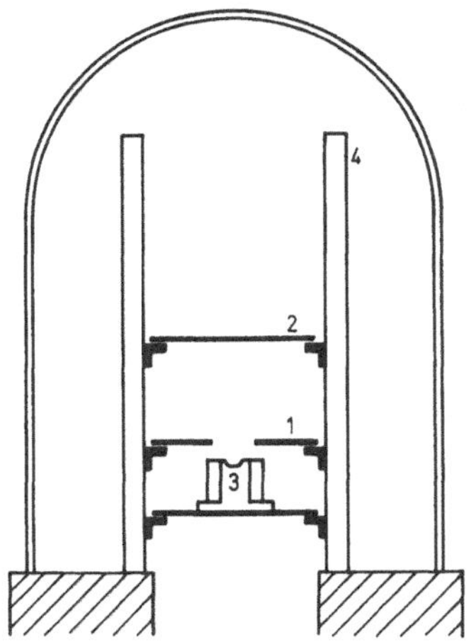

Fig. 1. Reactor for film deposition

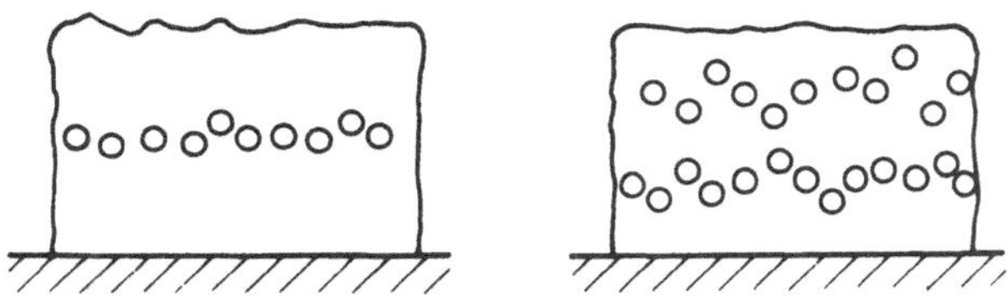

Fig. 2. Two-dimensional (left) and three-dimensional (right) metal particle distribution in the polymer matrix

factor were deposited on the lower side of the upper electrode. Films with spatially varying filling factor were obtained on the upper side of the lower electrode.

It is possible to prepare plasma polymer metal composite films with two- or three-dimensional metal particle distribution in the polymer matrix (Fig. 2). Two-dimensional particle distribution is obtained by step-by-step plasma polymerization and metal evaporation, or simultaneous plasma polymerization and metal evaporation for a short period. A three-dimensional distribution of the metal particles in the polymer matrix has been achieved by simultaneous plasma polymerization for a longer time and increased monomer flow rate, as well as by more than one metal evaporation interrupted by plasma polymerization steps.

3. Structure

There are three general structural regions for thin composite films consisting of metal-insulator mixtures [9]. At first, the metal film may be closed and polymer inclusions can be found. Therefore, the filling factor must be higher than the percolation threshold f_c. Near this threshold the films show the behavior characteristic for the percolation region. Now, there is no longer a closed metal layer, and the formation of metal clusters has been started, but these clusters are not completely separated from each other. The third structural region is characterized by a polymer matrix and the metal particles are completely isolated from each other.

Structural investigations were carried out with a JEM 100 CX transmission electron microscope. For this purpose the films were deposited on KBr-predeposited glass substrates, lifted from the substrate in distilled water and picked up with the usual copper grids.

The microdiffraction pattern demonstrates that polycrystalline silver or gold was encapsulated in the composite film. Especially for composite films with high filling factors, the reflection intensity of the diffraction rings due to the metallic component is good. At lower filling factors the diffuse diffraction pattern of the polymer matrix is also obtained, comparable in intensity with the diffraction intensity of the metal clusters. In particular, at very thin plasma polymer silver composite films ($d < 30$ nm) it is probable that some silver particles are not completely encapsulated in the polymer matrix. After storage in air for 6 weeks, weak Ag_2S reflections were observed. After 18 months strong Ag_2S reflections were found. At thicker plasmapolymer silver composite films ($d > 100$ nm) no changes in the direction of corrosive products of metals were found after more than 2 years.

Plasmapolymer metal composite films could be prepared for all the above-mentioned structural regions, and it is possible to get all the structural regions at one film with spatially varying filling factor. Figure 3 shows four transmission electron microscope (TEM) micrographs of one plasma polymer silver composite film. The filling factor decreases continuously from Fig. 3 (a) to (d). Figure 3 (a) represents the metallic region of the composite. The sample of Fig. 3 (b) has a structure near to the percolation threshold. Figure 3 (c) and (d) show silver particles encapsulated in the polymer matrix. The silver particle size and the deviations from the spherical shape decrease with decreasing filling factor. At low filling factors the silver particles are spherically shaped.

Microdiffraction pattern of plasma polymer germanium composite films show that germanium is

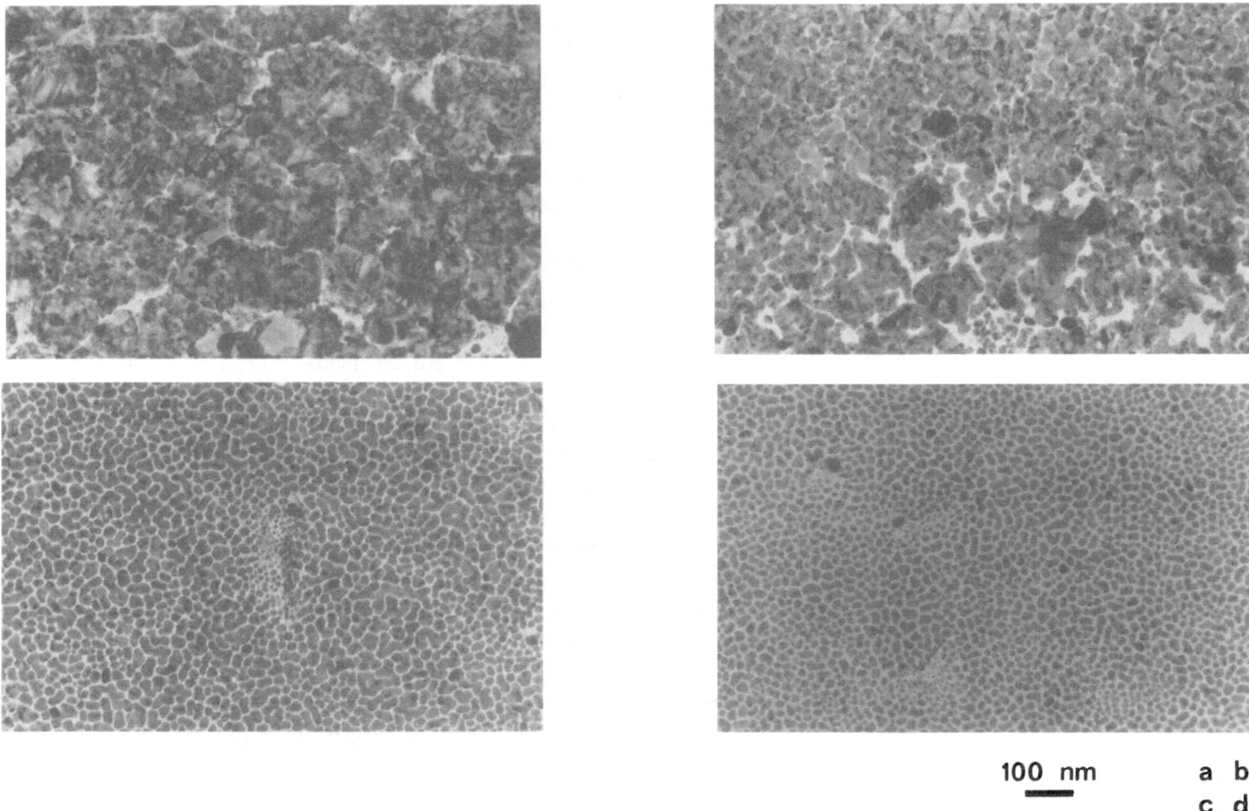

Fig. 3. TEM-micrographs of a plasma polymer silver composite film with spatially varying filling factor, thickness 60 nm. The filling factor decreases from (a) to (f)

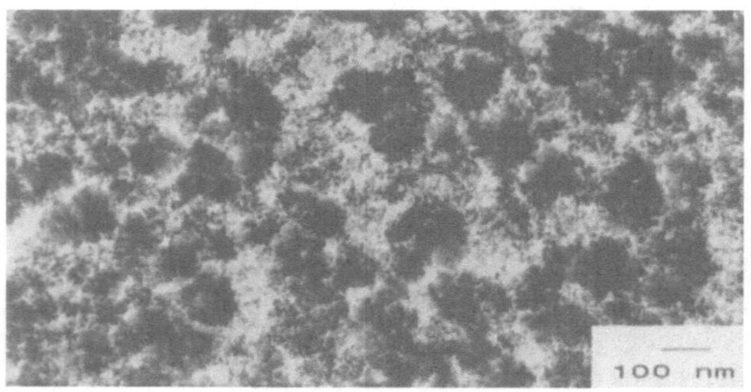

Fig. 4. TEM-micrograph of a plasma polymer germanium composite film, thickness 150 nm

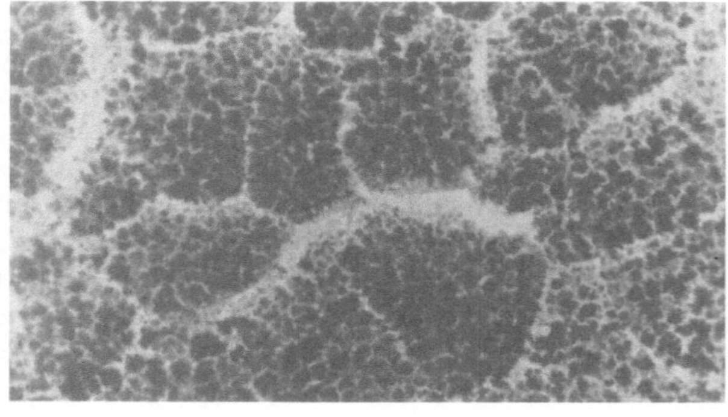

Fig. 5. TEM-micrograph of a plasma polymer holmium composite film, thickness 100 nm

encapsulated as amorphous material or as small crystallites depending on preparation parameters. The TEM micrograph (Fig. 4) shows that there are no sharp borders between the polymer matrix and the germanium clusters. Amorphous material and crystallites can also be found on the same sample. Plasmapolymer holmium composite films contain only amorphous metal clusters. Figure 5 gives a TEM micrograph of such a film. The amorphous material decorates the so-called cauliflower morphology [10] of the plasma polymer. The occurrence of this structure depends on the preparation properties, and it could be already found at low thickness.

4. Electrical properties

The electrical properties of metal insulator composite films are different for different structural regions of the composites. At the metallic region, the conductivity is similar to thin metal films and can be described with the limitation of the mean free path of the conduction electrons. In the percolation region, the conductivity decreases some orders of magnitude, and percolation theories [11] can be used for modeling the electrical behavior. At the insulating region, the conductivity is low and the films show insulator behavior. For the description of the conduction mechanism mostly thermal-activated tunneling between the metal islands was assumed [12].

The d.c. conductivity of plasma polymer metal composite films was studied by using lateral vacuum evaporated metal electrodes on glass substrates. Composite films with spatially varying filling factor were deposited on these electrodes. The d.c. conductivity was calculated based on the measured current-voltage relationship, and on the slit geometry, as well as on the film thickness.

The occurrence of different conduction mechanisms due to the different structural regions is typical for plasma polymer metal composite films with continuously varying filling factors. Figure 6 shows seven examples of the current voltage relationship of one plasma polymer silver composite film. Sample XI has the highest and sample V the lowest filling factor. The samples XI and X represent the metallic structural region and the conductivity is about 10^3 ... 10^4 Ω^{-1} cm^{-1}. In the insulating region the samples V to VII show low conductivities of 10^{-8} ... 10^{-7} Ω^{-1} cm^{-1}. This is substantially higher than the conductivity of the pure plasmapolymer of about 10^{-15} Ω^{-1} cm^{-1} [13]. The samples VIII and IX belong to the percolation region. The conductivity is 10^{-5} ... 10^{-4} Ω^{-1} cm^{-1} for sample VIII and 10^{-1} ... 10 Ω^{-1} cm^{-1} for sample IX, but hysteresis effects exist and also switching behavior could be observed. Some samples show current switching from 10^{-8} to 10^{-4} A at a voltage of 8 V. Reswitching took place at 0.1 V depending on preceding switching processes. Switching and reswitching should be connected with microstructural changes.

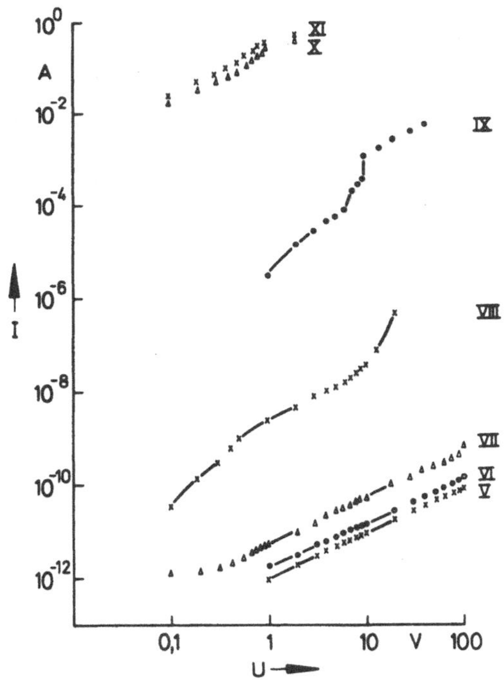

Fig. 6. Current-voltage relationship of seven samples of a plasma polymer silver composite film, thickness 70 nm. The filling factor decreases from slit XI to slit V

5. Optical properties

Pure plasma polymer thin films show an increasing optical transmission from the UV to the NIR range [14]. Especially in the UV region the optical transmission depends strongly on the film thickness and on the monomer used for polymerization. In the visible range the transparency is higher than 90% for films smaller than 1 μm. At thin discontinuous metal films an optical absorption caused by collective plasma oscillations could be observed [15]. At silver or gold particles these plasma resonance absorptions were found as transmission minima in the visible spectral range, also when the particles are encapsulated in an insulating matrix [16].

For our optical investigations, we deposited plasma polymer metal composite films with spatially varying filling factor on quartz substrates and used conventional spectrophotometers.

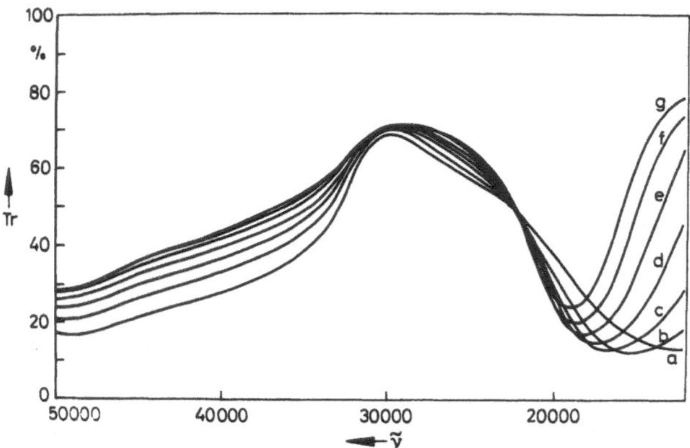

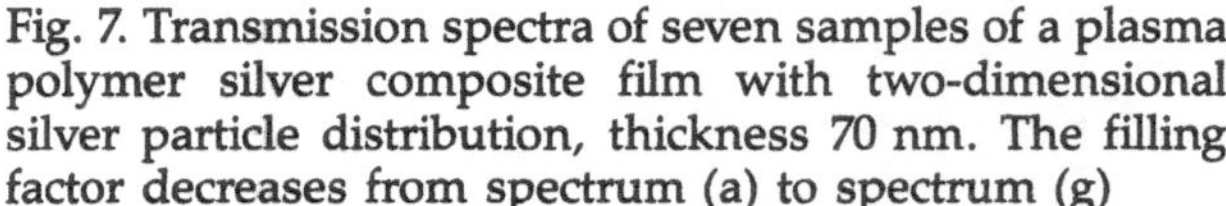

Fig. 7. Transmission spectra of seven samples of a plasma polymer silver composite film with two-dimensional silver particle distribution, thickness 70 nm. The filling factor decreases from spectrum (a) to spectrum (g)

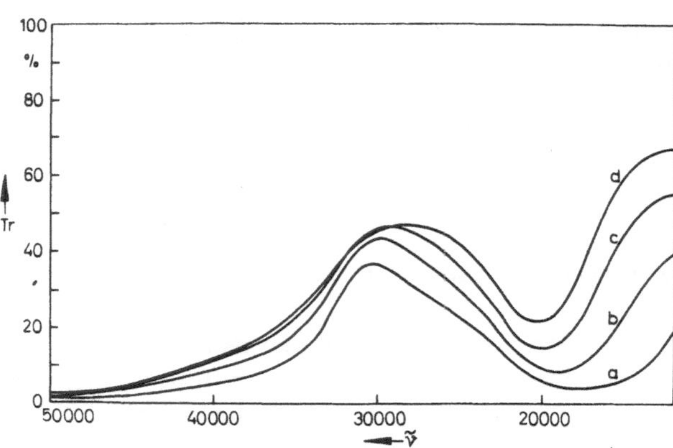

Fig. 8. Transmission spectra of four samples of a plasma polymer silver composite film with three-dimensional silver particle distribution, thickness 90 nm. The filling factor decreases from spectrum (a) to spectrum (d)

Figure 7 shows transmission spectra of a 100-nm-thick plasma polymer silver composite film with two-dimensional particle distribution in the polymer matrix. The filling factor decreases continuously from spectrum (a) to spectrum (g). With decreasing filling factor plasma resonance absorption shifts in the direction to higher wavenumbers. The width of the resulting transmission minima also decreases. It should be emphasized that the spectra show a constant transmission point at 22000 cm^{-1} over a broad filling factor interval. At this wavenumber the optical transmission is independent on the particle size and on particle shape.

The optical properties of plasma polymer silver composite films with three-dimensional particle distribution in the polymer matrix (Fig. 8) differs essentially from the properties of films with quasi two-dimensional particle distribution. The filling factor decreases continuously from spectrum (a) to spectrum (d) and it is generally lower than for the film shown in Fig. 7 the shift of the transmission minima is small.

For the description of the optical properties of plasma polymer metal composite films effective medium theories can be used. The properties of the two basic materials, e.g., the dielectric funtions of the metal and the insulator were connected to an effective property. There are different parameters, e.g., filling factor and depolarization factor representing the microstructure of the composite film. Detailed discussions and calculations of the optical properties of plasma polymer silver composite films compared with the experimental data can be found in [14].

6. Structure modifications

The microstructure of plasma polymer metal composite films substantially changes by thermal annealing procedures. Figure 9 shows TEM-pictures of three samples of one plasma polymer silver composite film before (left) and after (right) the thermal annealing at 450 K using the in situ heating unit of the electron microscope. The obtained drastic changes of particle size and particle shape were designated as coalescence processes. One coalescence condition is that the silver particles are not very strongly fixed inside of the polymer matrix. The coaelscence process involves two steps:

1) Two or more particles coalesce to a larger one.
2) Irregular, nonspherical particles become more spherical.

At sample (a) the structure of the film has changed from the metallic region to the insulator region and a small number of large particles was obtained. For the sample near the percolation threshold (c) and the sample of the insulating region (e) drastic microstructural changes have also occurred. During the coaelscence the percolation threshold f_c shifts to a higher value.

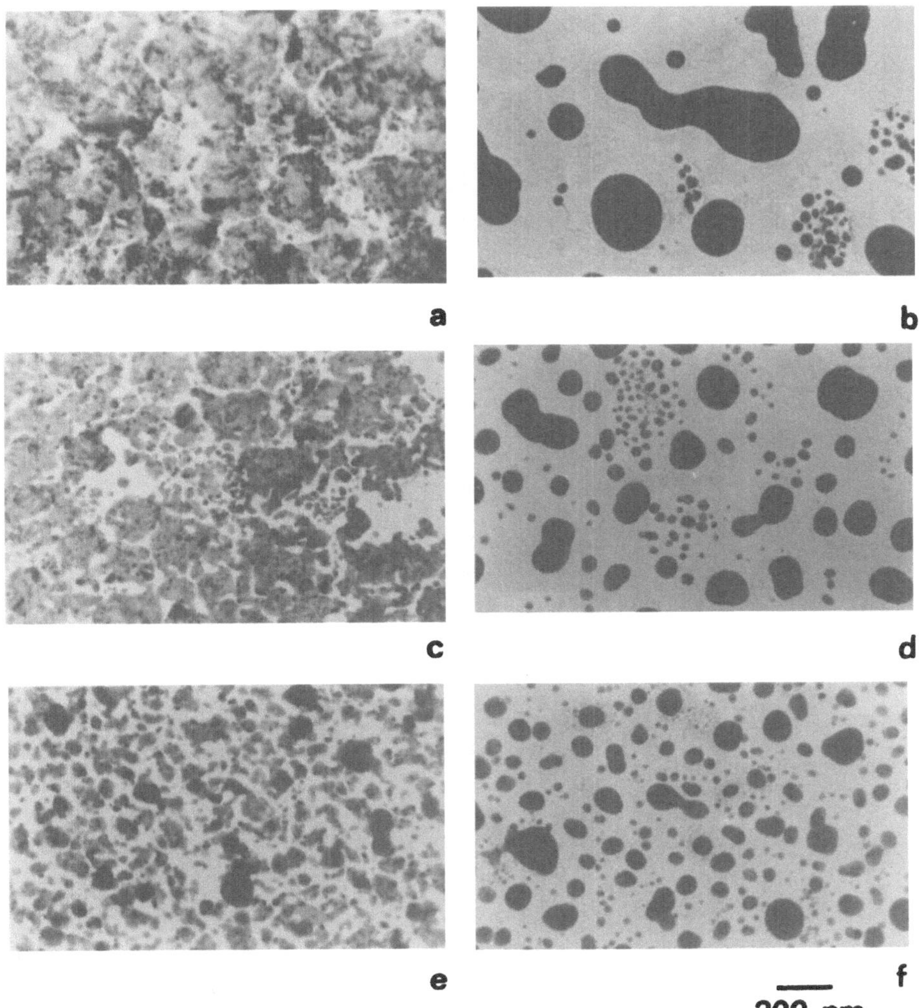

Fig. 9. TEM-micrographs of three samples of a plasma polymer silver composite film, thickness 60 nm. The micrographs on the left side show the sample before, and on the right side after the in situ annealing up to 450 K

These microstructural changes have essential influence on the electrical properties of the films. Figure 10 shows the current temperature relationship of four samples of the insulating region of a plasma polymer silver composite film. After a weak increase of the current due to the thermal activation of the electron tunneling at about 370 K an essential decrease of the conductivity starts. It can be assumed that the coalescence processes start at this temperature. At 480 K the conductivity of the samples is substantially lower. The tunneling current depends mainly on the average distance between two neighboring particles, and after coalescence this distance is greater (see Fig. 9). At further annealing procedures up to 480 K the current is lower, but a small increase of the current depending on temperature could be observed as before.

The thermal coalescence can be prevented by previous electron irradiation of the sample in the electron microscope. The irradiated area and the irradiation intensity were controlled by means of the goniometer and the condensor lens. Figure 11 shows a plasma polymer silver composite film after in situ annealing to 470 K. Before annealing the film was irradiated with 100 keV electrons at 10^3 pA/cm^2 (dark region) or 10^2 pA/cm^2 (light region). In the dark region no changes of the microstructure during annealing were observed. To the contrary, in the light region coalescence took place. During electron irradiation the plasma polymer in the dark region was hardened, and then the stability of the polymer matrix is sufficient to prevent the silver particle coalescence. Consequently, it is possible to limit the coalescence process to the nonirradiated areas.

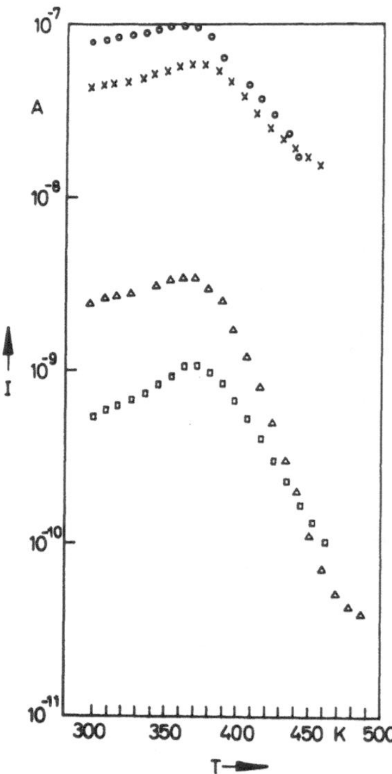

Fig. 10. Current temperature relationship of four samples of a plasma polymer silver composite film, thickness 70 nm. The samples were heated at 10^{-4} Pa with a heating rate of 4 K per min from room temperature to 470 K

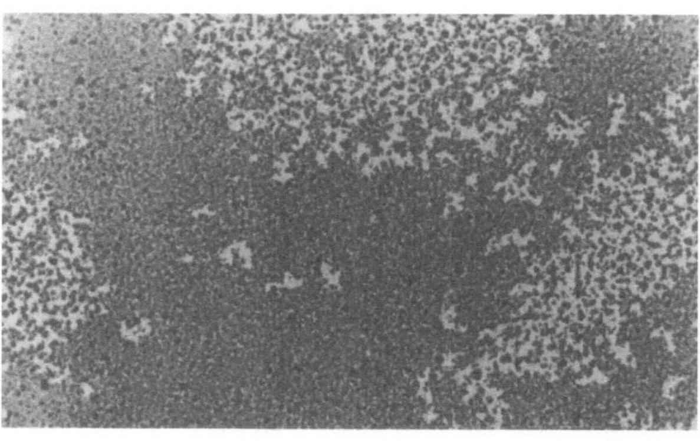

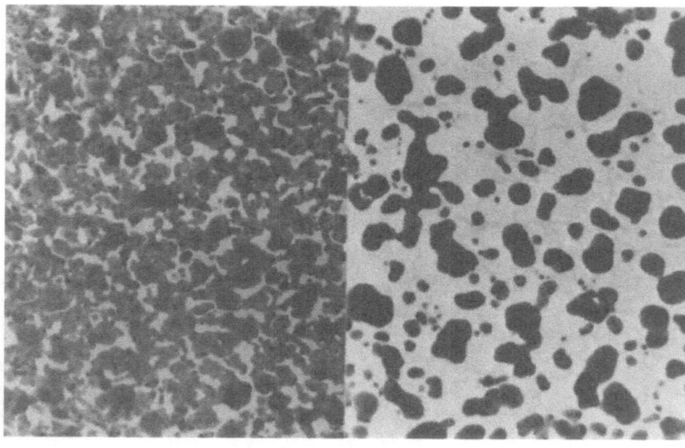

Fig. 11. TEM-micrographs of a plasma polymer silver composite film after in situ annealing up to 450 K, thickness 60 nm. The upper part shows regions without structural changes (dark) and regions with coalesced clusters (light). Dark region: 10 s irradiation with 100 keV electrons, 10^3 pA/cm^2. Light region: 10 s irradiation with 100 keV electrons, 10^2 pA/cm^2. The lower part shows regions without (left) and with coalescence (right) at higher magnification

In plasma polymer germanium composite films, holographic recording using a ruby laser (694 nm) was successful [17]. The reached sensitivity depends on the film thickness and on the surface morphology. The textured surface (see the scanning electron microscope (SEM) picture in Fig. 12 (a)) can be melted, and laser ablation took place. The trenches in Fig. 12 (a) marked by the arrows are the result of the laser irradiation. Figure 12 (b) shows one of these trenches in normal incidence, and with higher magnification. Up to and below this trench there are no changes of the original textured surface. On 1-μm-thick plasmapolymer germanium composite films a diffraction effectivity of 10% and a sensitivity of 0.005 J/cm^2 was reached.

7. Conclusions

The great variety of composite films achieved by using different metals shows that the combination of plasma polymerization and metal evaporation is a simple and successful method to prepare plasma polymer metal composite films. Electrical and optical properties depend essentially on the microstructure of the films. Thermal annealing of the films causes substantial microstructural changes due to coalescence processes. There is a strong decrease of the electrical conductivity from 10^2 to 10^{-8} Ω^{-1} cm^{-1} for samples with an order near to the percolation theshold. By using small parts of the film, we succeeded in preventing coalescence.

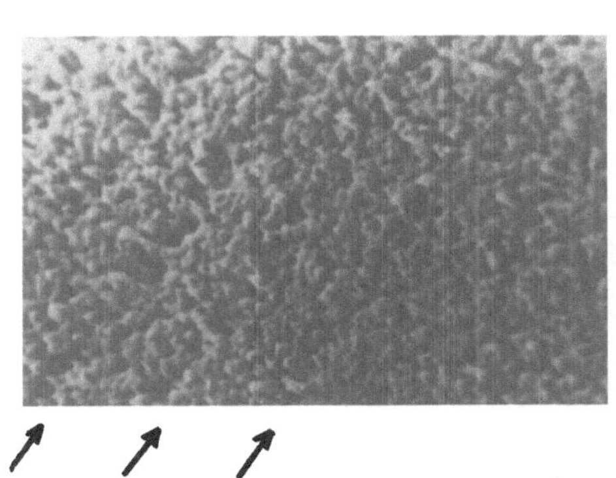

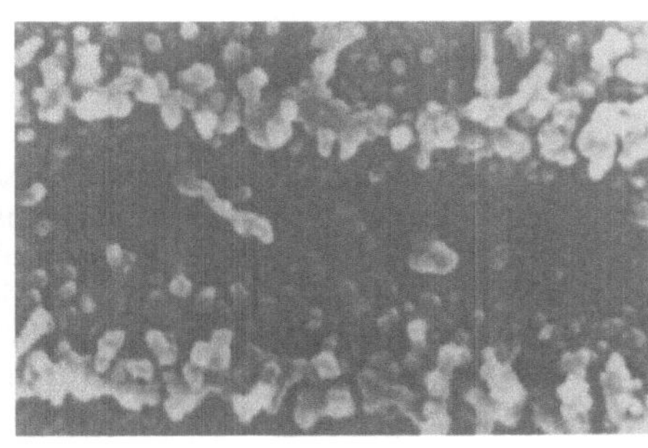

Fig. 12. Surface topography observed by SEM of a 150-nm-thick plasma polymer germanium composite film after holographic recording [17]: a) incidence direction 60°; b) normal incidence. The arrows on the lefthand picture mark trenches created by laser ablation. The righthand picture shows that inside such a trench there is no longer a textured surface

By laser annealing, it was possible to create a new film morphology and to store information in special composite films. Plasma polymer germanium composite films can be used for holographic recording.

Acknowledgements

The authors thank Do Ngoc Uan for his assistance with the structure investigations.

References

1. Hamann C, Kapfrath G (1984) Vacuum 34:1053
2. Kay E (1986) Z Phys D 3:251
3. Dilks A, Kay E (1979) ACS Symp Ser 108:3
4. Kay E, Dilks A, Hetzler U (1978) J Macromol Sci Chem A 12:1393
5. Martinu L, Biederman H, Zemek J (1985) Vacuum 35:171
6. Kashiwagi K, Yoshida Y, Murayama Y (1987) J Vac Sci Technol A 5:1828
7. Sadhir RK, Saunders HE (1985) J Vac Sci Technol A 3:2093
8. Morosoff N, Patel DL, Ehite AR, Umana M, Crumbliss AL, Lugg PS, Brown DB (1984) Thin Solid Films 117:33
9. Abeles B, Sheng P, Coutts MD, Arie Y (1975) Adv Phys 24:407
10. Heilmann A, Hamann C, Kampfrath G, Do Ngoc Uan (1989) Phys stat sol a 114:551
11. Stauffer D (1979) Phys Rep 54:1
12. Morris JE, Coutts TJ (1977) Thin Solid Films 47:3
13. Kampfrath G, Duschl D, Hamann C (1987) Acta Polym 38:389
14. Heilmann A, Kampfrath G, Hopfe V (1988) J Phys 21:986
15. Kreibig U, Genzel L (1985) Surf Sci 156:678
16. Cohen RW, Cody GD, Coutts MD, Abeles B (1973) Phys Rev B 8:3689
17. Savchuk AV, Salkova EN, Sergan TA, Heilmann A (1990) Phys stat sol a 122:K83

Received January 19, 1991
accepted February 8, 1991

Authors' address:

A. Heilmann
Universität of Technology Chemnitz
Department of Physics
P.O. Box 964
O-9010 Chemnitz, FRG

Discussion

HECKNER:
What is the physical model for photocurrents in the plasma polymer metal films? A possibility should be Schottky contacts between metal clusters and surrounding polymer.

HEILMANN:
There are first measurements of the photocurrent of plasma polymer silver composite films on slit electrodes. The photocurrent show a significant spectral dependence and it occurs only in a special structural region near to the

percolation threshold. There are different models under discussion, and Schottky contact is one of them.

WENDORFF:

The material considered here has a disadvantage as a holographic storage medium in that it leads to an amplitude object, rather than to a phase object. Did you look at nonlinear optical properties of these composite systems, such as electro-optical properties? There are indications in the literature that they should display interesting nonlinear optical properties.

HEILMANN:

Such measurements are in preparation.

KREMER:

At the beginning of your talk, you presented conductivity measurements exhibiting a non-ohmic (nonlinear) behavior. Therefore, I wonder if you expect the effective medium theories to be applicable in your samples, because it is a linear approach.

HEILMANN:

We have used effective medium theories for description of the optical properties of films with filling factors below the percolation threshold. Effective medium theories are also useful to describe the electrical properties, especially the bounds as described by Wiener and Hashin/Shtrikman and by the Bruggeman theory. But there is only some information about the range of the conductivity in dependence on the filling factor. There is no information about the conductivity mechanism. For films with nonlinear electrical behavior the discussion mostly includes the possibility of microstructural changes, e.g., conducting "channels" during the electrical measurements.

Preparation and characterization of glassy plasma polymer membranes

J. Weichart and J. Müller

TU-Hamburg-Harburg, FSP 4—07, Hamburg, FRG

Abstract: Plasma chemical vapor deposition of hexamethyldisilazane was investigated for its applicability as solubility controlled gas separation membranes. Substrates were porous inorganic membranes (Anotec: 0.02 μ) for permeation measurements and silicon wafers for deposition rate, density, and infrared measurements. To ensure a uniform membrane on the porous support a deposit of five times the pore diameter was required. Membranes deposited at 2.45 GHz from HMDSN/O_2 mixtures exhibit a high deposition rate (1.8 nm · s^{-1}) and a wide density range (1.15—1.60 g · cm^{-3}). — Typical nitrogen permeation rate of a 100 nm film was 2 · 10^{-8} m^3 (STP) s^{-1} m^{-2} Pa^{-1}. Permeation rates of CO_2 and C_4H_{10} related to N_2 and the thermal dependence indicate that the permeation is viscosity controlled. These films exhibit a microgel-like structure with an estimated pore diameter of 2 nm. Anodic films prepared in a 13.56 MHz parallel plate reactor at low deposition rate (0.3 nm · s^{-1}) have densities between 1.3 and 1.5 g · cm^{-3}. They exhibit an infrared absorption of the δ:CH_3 vibration at ~1260 cm^{-1}, which is an easy accessible indication of crosslinking and oxidation of more than twice than in microwave plasma films. For an 0.6-μm anodic film an ideal separation factor for CO_2/N_2 and C_4H_{10}/N_2 of ~6 is obtained, which is comparable to that of 1-μm conventional polydimethylsiloxane.

Key words: Membrane; plasma polymerization; gas permeation; silicone

1. Introduction

Gas separation by membranes has become an interesting alternative to thermal separation methods. High permeation rates are achieved by composite membranes, where thin polymer films are deposited on a porous mechanical support. Conventional polymerization techniques are limited in producing ultrathin films which are free of pinholes. Therefore, plasma deposition from gas phase has been discussed [1].

Two types of separation can be utilized:

a) pore-size controlled diffusion (molecular sieve); and
b) solubility-controlled diffusion.

This study investigates if plasma silicone polymer films can compete with conventional silicones in a solubility-controlled separation. Films deposited at 13.56 MHz will be compared to microwave plasma films prepared under electron cyclotron resonance conditions (ECR, 2.45 GHz). It has been demonstrated that efficiency in microwave plasma-CVD-processes is very high [2] and that coating of three-dimensional substrates is possible [3]. These features are of interest in technical coating problems like the production of membranes. To exclude substrate effects the plasma polymer films were deposited on porous inorganic membranes.

2. Preparation

Hexamethyldisilazane (HMDS-N) has been chosen as monomer for the silicone-like membranes. A needle valve is used to dose the monomer vapor from a liquid reservoir at room temperature. The deposition processes were carried out in two HF-plasma-CVD-vessels:

a) At 13.56 MHz in a parallel plate reactor at a variable power input of 5 to 1000 Watt, at a process pressure of 5 Pa, and varying oxygen partial pressure. This apparatus has been described elsewhere [4].
b) In a microwave plasma-CVD-apparatus (2.45 GHz) at electron cyclotron resonance (ECR) (Fig. 1). In this reactor the microwave power is fed via a helix antenna into a quartz tube with a gas supply and permanent SmCo-magnets at one end. The sample is situated at the other end of the tube, where reaction gases are pumped out via slits. It this downstream position the sample is nearly free of thermal heating effects by the plasma. Typical monomer flow rates were 2.9 and 6.7 mmol HMDS-N per h. Oxygen was introduced by a mass flow controller at a maximum rate of 5 sccm/min (12.2 mmol/h). The reactor was evacuated to $3 \cdot 10^{-4}$ Pa before introducing the gas. Its maximum leak rate was 0.02 mmol O_2/h. Process pressure was measured by a MKS Baratron, which takes values between 0.4 and 1.3 Pa.

The silicone-like layers were deposited on silicon monitor wafers for gravimetric, ellipsometric, and IR-transmission measurements, and on porous Al_2O_3 membranes (Anodisc 47, 0.02 μ, Anotec, supplied by Merck) exhibiting a large open area with a regular pore diameter of 20 nm.

3. Characterization

The film weight was mesured with a microbalance (Sartorius Research R 200 D). Film thickness and index of refraction were measured by ellipsometry (Gaertner L 116 A). From these measurements the film density could be calculated within ± 0.02 g/cm^3.

3.1 Infrared spectroscopy

Infrared measurements were carried out on a double beam Fourier transform IR-spectrometer (Perkin Elmer 1830). Silicon monitor wafers were used as substrates for transmission measurements. Each substrate was recorded before film deposition as reference, so that difference spectra represented the film alone. With the film thickness determined by ellipsometry, the absorbance coefficients could be calculated. The HMDSN-plasma-deposits absorb at the characteristic wavenumbers summarized in Table 1.

3.2 Gas permeation

For the investigation of gas permeation, pieces of Anodisc-membranes were mounted on bored metal gaskets. They were coated in the plasma chamber together with a silicon slice for thickness control and mounted in the permeation cell (Fig. 2). For the measurements, at first, both sides of the membrane

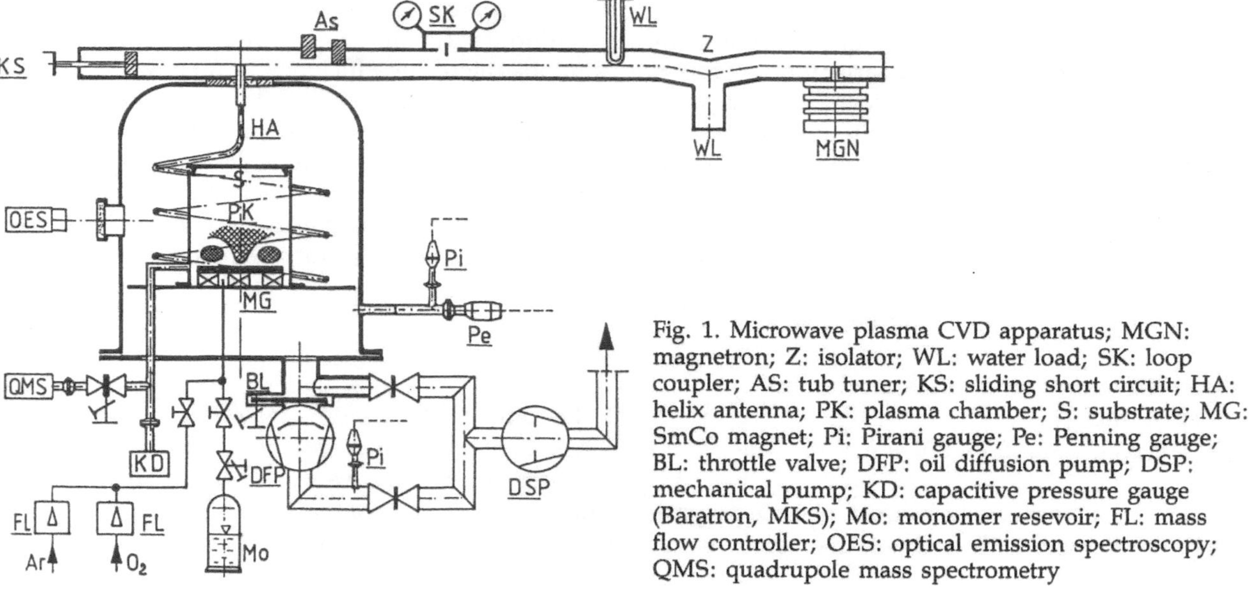

Fig. 1. Microwave plasma CVD apparatus; MGN: magnetron; Z: isolator; WL: water load; SK: loop coupler; AS: tub tuner; KS: sliding short circuit; HA: helix antenna; PK: plasma chamber; S: substrate; MG: SmCo magnet; Pi: Pirani gauge; Pe: Penning gauge; BL: throttle valve; DFP: oil diffusion pump; DSP: mechanical pump; KD: capacitive pressure gauge (Baratron, MKS); Mo: monomer resevoir; FL: mass flow controller; OES: optical emission spectroscopy; QMS: quadrupole mass spectrometry

Table 1. Infrared absorption band observed in HMDSN/O_2-plasma polymer films

3400—3300 cm^{-1}:	ν: OH, stretch vibration;
3000—2800 cm^{-1}:	ν: CH, CH_2, CH_3, stretch vibration modes;
2300—2100 cm^{-1}:	ν: SiH, stretch vibration;
1276—1253 cm^{-1}:	δ_s: CH_3, symmetric deformation in $Si(CH_3)$;
1090—1020 cm^{-1}:	ω: Si—CH_2—Si, wagging, and ω_{as}: Si—O—Si, asymmetric stretch;
950—900 cm^{-1}:	ω_{as}: Si—N—Si, asymmetric stretch;
850—840 cm^{-1}:	ρ: CH_3 rocking in $Si(CH_3)_3$;
810—800 cm^{-1}:	ω: Si—C, stretch vibration, and ρ: CH_3 rocking in $Si(CH_3)_2$, and Si—O—Si, bending

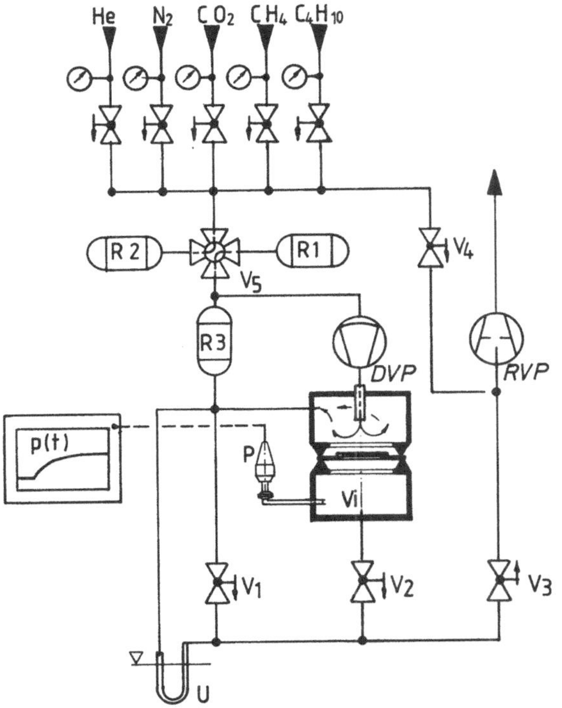

Fig. 2. Permeation cell: RVP: sliding vane rotary vacuum pump; DVP: diaphragm vacuum pump; R1, R2, R3: reservoirs; V_i: closed volume; P: pressure gauge (Baratron, MKS); V1 to V5: valves; U: Hg-difference manometer

were evacuated by a rotary pump; then they were refilled with the selected gas from the gas manifold to a pressure of 800 mbar. V2 was closed, and by switching V5, a defined volume of reservoir R1 or R2 was introduced to a pressure difference of about 200 mbar between both sides of the membrane. A continuous gas flow and constant pressure is guaranteed by a diaphragm pump and the reservoir R3.

The permeating gas flux Q is proportional to the difference pressure Δp:

$$Q = P \cdot \frac{F}{l} \cdot \Delta p \,, \tag{1}$$

with F the membrane area, l the membrane thickness, and P the permeation coefficient.

The pressure rise in the lower chamber is measured by a capacitive membrane gauge P (Baratron, MKS). The rise is characterized by a time constant τ in:

$$p = p_0 + \Delta p_0 \cdot (1 - \exp(-\tau t)) \,. \tag{2}$$

Knowing τ, the lower chamber volume V_i, and the membrane area F, the permeation rate coefficient $P^* = P/l$ can be calculated.

The film thickness l is measured on a sliced silicon wafer, coated in the same deposition run. Assuming that the film thickness on the porous support is equal to that on a flat one, the permeation coefficient P can be calculated. The ideal separation factor of two gases, A and B, is defined as the ratio of the determined single gas permeation rates:

$$\alpha(A/B) = P(A)/P(B) \,. \tag{3}$$

In Fig. 3 the permeation rate for nitrogen $P^*_{N_2}$ of a HMDSN-O_2-equimolar membrane is plotted vs the film thickness. Evidently, the support pores were covered with a continuous layer at a thickness of 100 nm, which is five times the pore diameter.

4. Results

4.1 Deposition at 13.56 MHz

The deposition rate on the anode of the parallel plate reactor is shown as function of plasma power in Fig. 4. It exhibits a plateau between 10 and 50 Watts. The deposition rate is increasing at increasing oxygen partial pressure. The film density (Fig. 5) is at maximum in the power range of

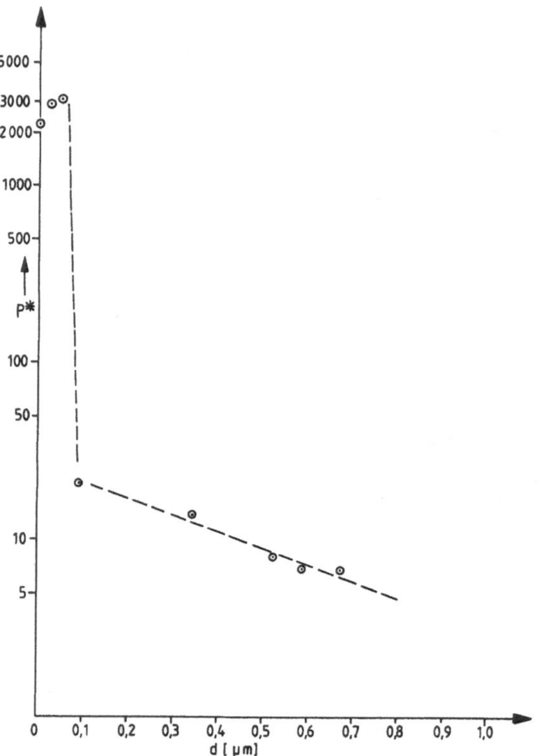

Fig. 3. Nitrogen permeation in [$\cdot 10^{-9}$ m^3 (STP) $\cdot$ m^{-2} $\cdot$ s^{-1} $\cdot$ Pa^{-1}] of Anodisc membranes coated by microwave CVD of HMDS-N/O_2 as function of film thickness, measured on a silicon wafer

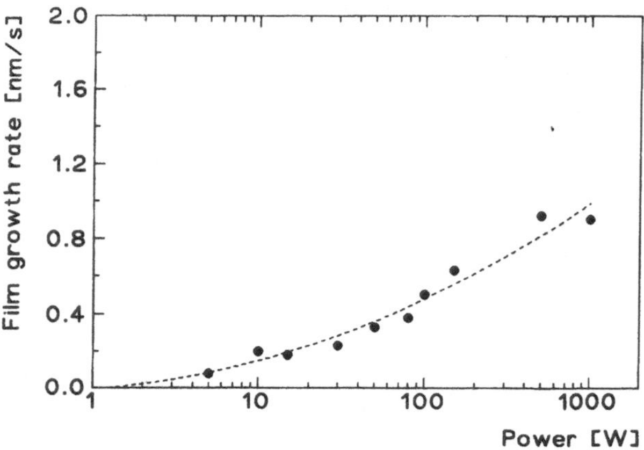

Fig. 4. Film growth rate of 13.6 MHz anodic deposition in [nm $\cdot$ s^{-1}] as function of plasma power [W] (HMDSN: 25 mmol/h; O_2: 67 mmol/h)

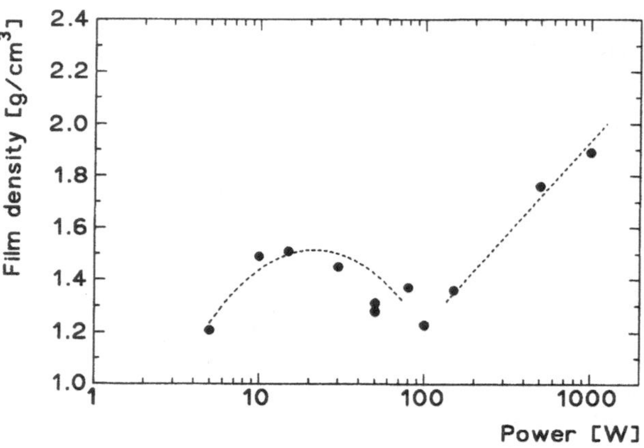

Fig. 5. Film density of 13.6 MHz anodic deposition in [g $\cdot$ cm^{-3}] as function of plasma power [W] (HMDSN: 25 mmol/h; O_2: 67 mmol/h)

10—20 W. For layers deposited in this power range an ideal separation factor α (CO_2/N_2) of 5.8 was observed, if 70 mol% oxygen was added. This behavior vanishes as RF-power or oxygen partial pressure are increased.

4.2 Deposition at 2.45 GHz

For the microwave deposition process, the process pressure was kept below 3 Pa to prevent volume polymerization. Depending on plasma power the films deposited from pure HMDS-N vapor show densities between 1.1 and 1.4 g/cm^3 for low and medium microwave power input (50—130 W; 130—300 W resp.). It has to be remarked that power is measured in the transmitting waveguide and does not represent the real power in the plasma. In the high power region (300—500 W) volume polymerization was observed. By adding oxygen the deposition rate (Fig. 6) increases. Its maximum is observed at equimolar HMDS-N and O_2. At this gas mixture a reaction to silicone chains is likely:

$$[(CH_3)_3Si]_2NH + O_2$$

$$\Rightarrow [-O-Si(CH_3)_2-O-Si(CH_3)_2-] + N, C, H. \quad (4)$$

For higher oxygen flow rates the methyl groups were oxidized, resulting in a decrease of the deposition rate (Fig. 6). Simultaneously, the density

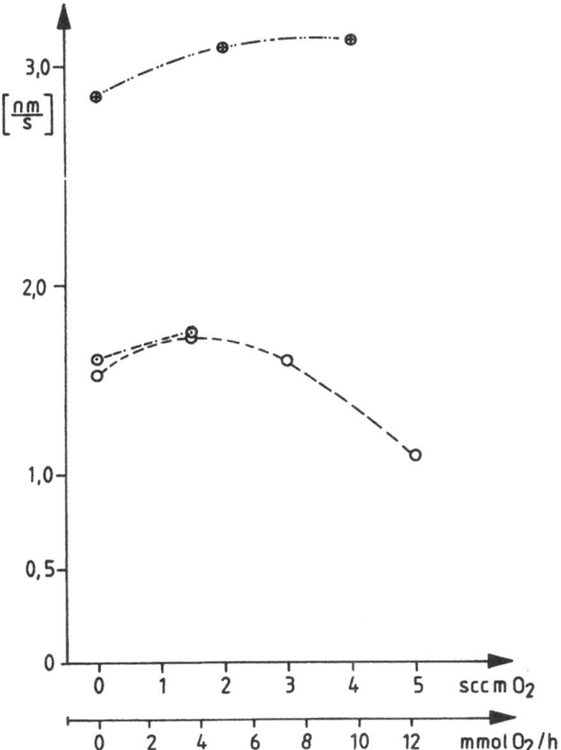

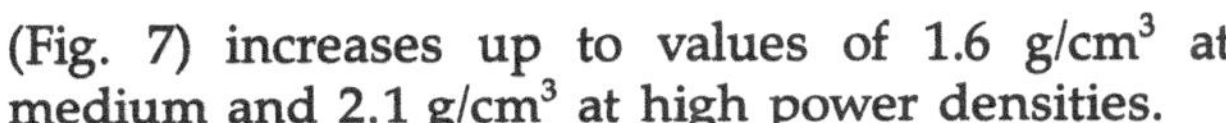

Fig. 6. Film growth-rate of ECR microwave plasma deposition in [nm · s^{-1}] as function of added oxygen in [sccm] and [mmol/h] for:
⊕ 6.7 mmol HMDS-N/h, 110 W; o 2.9 mmol HMDS-N/h, 110 W, deposition time = 300 s; ⊙ 2.9 mmol HMDS-N/h, 110 W, deposition time = 120 s

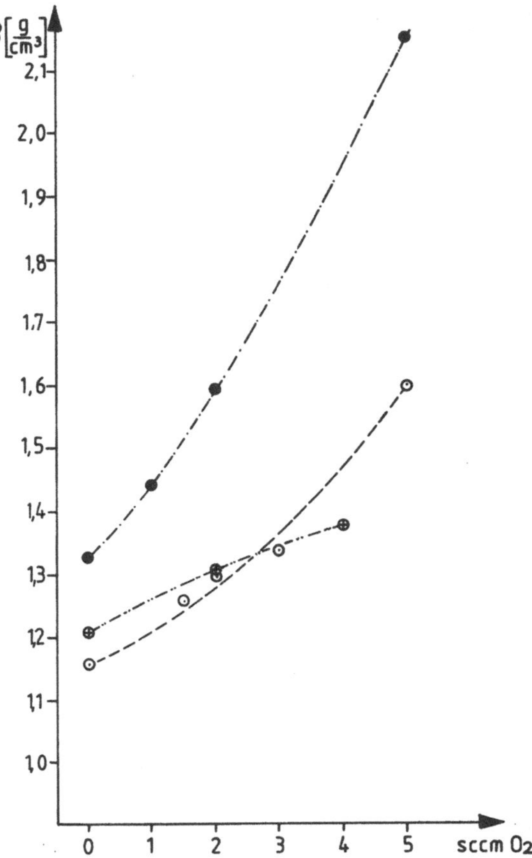

Fig. 7. Film density of ECR microwave plasma deposition in [g · cm^{-3}] as function of added oxygen in [sccm]: ⊕ 6.7 mmol HMDS-N/h, 110 W; ⊙ 2.9 mmol HMDS-N/h, 110 W, deposition time = 300 s; ● 2.9 mmol HMDS-N/h, 270 W

(Fig. 7) increases up to values of 1.6 g/cm^3 at medium and 2.1 g/cm^3 at high power densities.

4.3 *Comparison of film properties*

As shown in Table 2, the deposition rates obtained by microwave plasma CVD are remarkably higher than in the 13.6 MHz plasma; the film densities are comparable. For the deposition at 13.6 MHz, the maximum separation factor CO_2/N_2 does not correspond with the minimum in the film density, but with a local maximum of 1.4 g/cm^3. Figure 8 shows the infrared absorption spectra obtained for the two preparation methods. They obvisously differ at 950 cm^{-1}, where the microwave plasma films exhibit an asymmetric stretch vibration of Si—N—Si. A strong indication of crosslinking and oxidation is the position of the δ:$CH_3(SiCH_3)$-absorption, which shifts from 1253 to 1276 cm^{-1} with increasing oxygen concentration, applied power and deposition time, the latter resulting in stronger crosslinking by electron and UV-radiation.

It is difficult to quantitatively analyze the region between 1090 and 800 cm^{-1}, because of many interferences and broad peaks. Adding oxygen to the plasma ν_{as}:Si—O—Si becomes dominant, peaks between 1000 and 810 cm^{-1} vanish, accompanied by a decrease of ν:CH, ν:SiH and δ_s:CH_3; at the same time, absorption due to OH-groups at 3400 cm^{-1} increases rapidly. The methyl rocking vibrations at 841 cm^{-1} ($Si(CH_3)_3$) and 800 cm^{-1} ($Si(CH_3)_2$) can only be found for high flow rates and a low power input. A quantitative analysis of the δ:CH_3-absorption near 1260 cm^{-1} reveals an integral

Table 2. Nitrogen permeation coefficients, densities, deposition rates, and separation factors of films deposited on Anodisc membranes and references

Membrane preparation	Nitrogen permeation coefficient P in $\left[\cdot 10^{-15} \frac{m^3 \text{ STP } m}{m^2 \cdot s \cdot Pa}\right]$	Density [g/cm³]	Deposition rate [μm/h]	Ideal separation factors		
				α (CO_2/N_2)	α (C_4H_{10}/N_2)	α (He/N_2)
Microwave CVD HMSDN/O_2 2.9 mmol/h; 110 W	4.5	1.2	6.8	0.6	0.8	2.5
Microwave CVD HMSDN/O_2 6.7 mmol/h; 50 W	7.6	1.2	8.8	0.8	0.6	2.2
HF-CVD (13.6 MHz) HMDSN/O_2 anode, 30 W	0.85	1.45	0.83	5.8	6.4	6.1
HF-CVD (13.6 MHz) HMDSN/O_2 anode, 50 W	2.1	1.3	1.2	2.6	3.6	3.3
Anodisc, uncoated, 60 μ	—	—	—	0.8	0.8	2.7
Polydimethyl-siloxane (Wacker) on polyetherimide support	3.5	1.04	—	6.6	6.5	1.7

extinction of 2.5 to 4.2 A $cm^{-1}\mu m^{-1}$ for the films of the 13.6 MHz anodic deposition, against 0.9 to 1.5 A cm^{-1} μm^{-1} for the microwave plasma films.

The main results of permeation measurements are listed in Table 2. As a reference a 1-μm polydimethylsiloxane membrane (Wacker) on a porous polyetherimid support is included. The permeation coefficient P normalized to the film thickness is three times higher for polydimethylsiloxane than for plasma polymer films with solubility controlled diffusion.

Temperature dependent permeation measurements were carried out between 300 and 420 K to decide the type of permeation mechanism. For the solubility-controlled membranes prepared at 13.6 MHz, activated diffusion is observed, whereas microwave plasma films exhibit a weak temperature dependence, as is expected for microporous films.

5. Conclusions

It appears that high gas solubility like in silicone rubber can only be achieved under extremely soft plasma conditions, resulting in low deposition rates. The attempt to produce solubility-controlled membranes by microwave plasma CVD of HMDS-N has not yet been successful, although film density could be adjusted within a wide range and high amounts of organic groups could be incorporated, as can be seen from IR measurements.

Anodic plasma polymers prepared at 13.6 MHz exhibit elevated permeation rates for CO_2 and

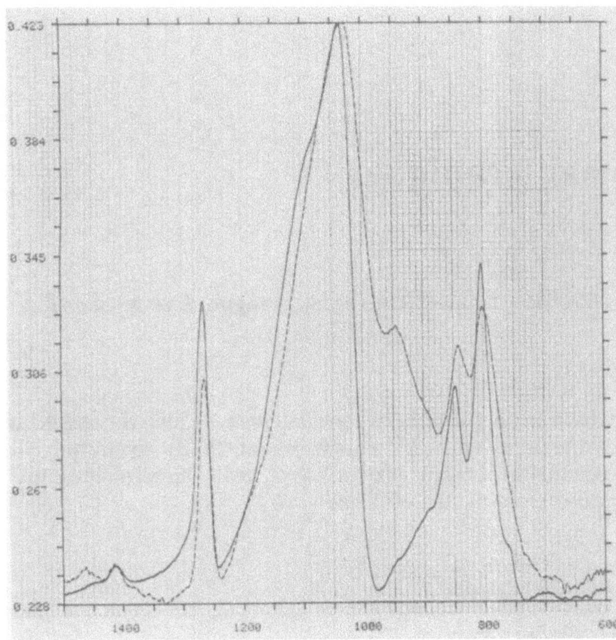

Fig. 8. Typical infrared absorption spectra between 1500 and 600 cm^{-1} of $HMDSN/O_2$ plasma polymer films prepared at: 13.6 MHz ——; 2.45 GHz, ECR – – –

C_4H_{10} due to an increased gas solubility in the membrane. Considering the low deposition rates and densities listed in Table 2, the deposition mechanism on the anode is probably governed by condensation and subsequent crosslinking by radicals, negative oxygen ions, and UV-radiation.

In contrast to this mechanism, under ECR-microwave conditions, an extensive volume fragmentation is assumed which results in a non-compact structure of the film: a microgel. By the difference in the permeation rates of coated and uncoated anodisc membranes, a pore diameter of the microwave plasma films of less than 2 nm can be estimated.

Appendix

Recently the microwave CVD apparatus has been modified, so that processes at lower power (30 W) and higher monomer flow rate (14 mmol/h) are possible. The deposition rate has decreased and is now about 0.4 μm/h. The corresponding films exhibit solubility controlled diffusion. The separation factor of microwave plasma CVD films: $CO_2/N_2 \approx 2$ is relatively poor compared with that of anodic deposition at 13.6 MHz. The density of these films remain below 1.25 g/cm^3 for oxygen portions up to 70 mol%. In 13.6 MHz anodic deposition the density is 1.45 g/cm^3 (Table 2), current membrane measurements yield a separation factor of $CO_2/N_2 \approx 15$.

References

1. Boenig H (1988) Fundamentals of Plasma Chemistry and Technology, Technomic Publishing Company, Inc., Lancaster, a review
2. Wertheimer MR, Moisan M (1985) Comparison of microwave and lower frequency plasmas for thin film deposition and etching. J Vac Sci Technol A3 6:1643
3. Weichart J, Müller J, Meyer B (1990) Beschichtung von dreidimensionalen Substraten mit einem magnetfeldunterstützten Mikrowellen-Plasma bei Raumtemperatur, Vakuum in der Praxis, 1/1991, Verlag Chemie, Weinheim
4. Peters D, Müller J, Sperling T (1990) Insulation and Passivation of 3-Dimensional Substrates by Plasma CVD Thin Films using Silicon-Organic Compounds. Second International Conference on Plasma Surface Engineering, 1990, Garmisch-Partenkirchen

Received January 4, 1991
accepted February 8, 1991

Authors' address:

J. Weichart
Technische Universität Hamburg-Harburg
FSP 4—07
Eissendorfer Straße 42
2100 Hamburg, FRG

Discussion

KREMER:

You showed an IR-spectrum of your plasma deposited films and observed an additional absorption band in the case of microwave (2.45 GHz) deposition in comparison to the HF (13.56 MHz) deposition. What is the physical reason for this difference?

WEICHART:

Although the composition of the supplied gas is nearly equal in both cases there are significant differences between the compared processes: Power input, pressure and flow rate are different, but the most important difference is the exciting frequency. At 13 MHz electron migration is possible, so that an asymmetric parallel plate reactor is polarized. The reaction mechanism in the plasma is not yet understood, but as this example shows, the incorporation of oxygen into the film is different. The absorption band at 950 cm^{-1} is due to nitrogen in the film. In the case of microwave plasma the deposition rate is ~10 times higher than in 13.6 MHz deposition, consequencing incomplete exchange of nitrogen by oxygen.

Basic physical properties of thin films prepared by unconventional techniques

H. Biedermann

Department of Polymer Physics, Faculty of Mathematics and Physics, Charles University, Prague, Czechoslovakia

Polymer thin films can be prepared by number of ways, however, several of them, which have been developed during the last decades, are using vacuum techniques. Mentioned methods are: 1) Plasma polymerization, 2) electron beam irradiation of the adsorbed organic vapours on the substrate, 3) u-v light irradiation of the absorbed organic vapours on the surface (substrate), 4) simple vacuum evaporation and 5) ICB technique. The first three methods have attracted considerable attention since 60-ties. Among them, plasma polymerization became the most popular. Basic physical properties of the obtained films are described and compared to the conventional polymers. Simple vacuum evaporation of polymer films, which is useful only in special cases, is also discussed in connection with ion-cluster-beam (ICB) technique. In conclusion future prospects of thin "unconventional" polymer films are outlined.

Author's address:

H. Biedermann
Charles University
Department of Polymer Physics
Faculty of Mathematics and Physics
Prague, Czechoslovakia

Various structure of plasma polymerized layers

A. Havránek and D. Slavínská

Department of Polymer Physics, Faculty of Mathematics and Physics, Charles University, Prague, Czechoslovakia

Abstract: The structure of plasma polymerized amorphous and crystalline layers is discussed. The discussion is based on literature data and on some of our experimental results obtained for polyvinylcarbazole and polyvinylalcohole layers. Optical and electron micrographs, and the surface profile graphs obtained by mechanical surfometer are the methods used for the detection of the plasma polymerized layer morphology. The structure of these layers is compared to that of ordinary polymers. It can be concluded that the plasma polymerized layers may be dealt with as polymer materials.

Key words: Plasma polymerized layers; crystalline structure; network; poly(N-vinylcarbazole (PVCa); roughness; surface profile

1. Introduction

The structure of plasma polymerized layers will be compared with that of classical polymeric materials. Various structural modifications of both types of materials are mentioned and their common features are emphasized. The aim of this contribution is to discuss the extent to which the layers prepared by plasma polymerization may be dealt with as polymeric materials. Results of other authors as well as some our experimental results will be used in the discussion.

2. Results and discussion

The structure of plasma polymerized layers is often assumed to be similar to the structure of plasma polymerized polyethylene (Fig. 1a) [1]. The polymeric nomenclature refers to it as to an irregular network. Many side chains occur in the network, and, especially, the presence of non-ethylene chemical units such as benzene rings and double bonds contribute to its irregularities.

Polymer networks are usually prepared by chemical cross-linking of linear polymers (vulcanization of natural rubber, as an example), by irradiation, or by polymerization of mers (chemical units which compose the polymer), functionality of which is higher than two. Irregularities occur even in these classical polymer networks. We shall give some examples. In vulcanized rubber, sulphur bridges are not only between two different linear macromolecules, but they also form loops beginning and ending at the same macromolecule (see Fig. 1b). Therefore, rings are present even in the most classical polymer network, not just in plasma polymerized polyethylene. The preparation of a polymer network by irradiation is, to some extent, similar to that performed by plasma polymerization. The linear polymer is destroyed by irradiation and bridges between macromolecules are created at the sites of destruction. Of course, this is an idealization. In fact, loops, radicals, side chains of limited length, and changes of the chemical structure of the units within the polymer chains are induced by irradiation similarly as by plasma polymerization. The structure of the network created during polymerization may be well described in the case of the so-called "model polymers", but the network is not simple even in this special case [2]. The defectness of networks created during polymerization from ordinary mers is large.

Thus, even as the proposed structure of plasma polymerized network has very many defects, it can be compared to the structure of polymer networks prepared in a usual manner. The irregularity of the networks implies their amorphous structure.

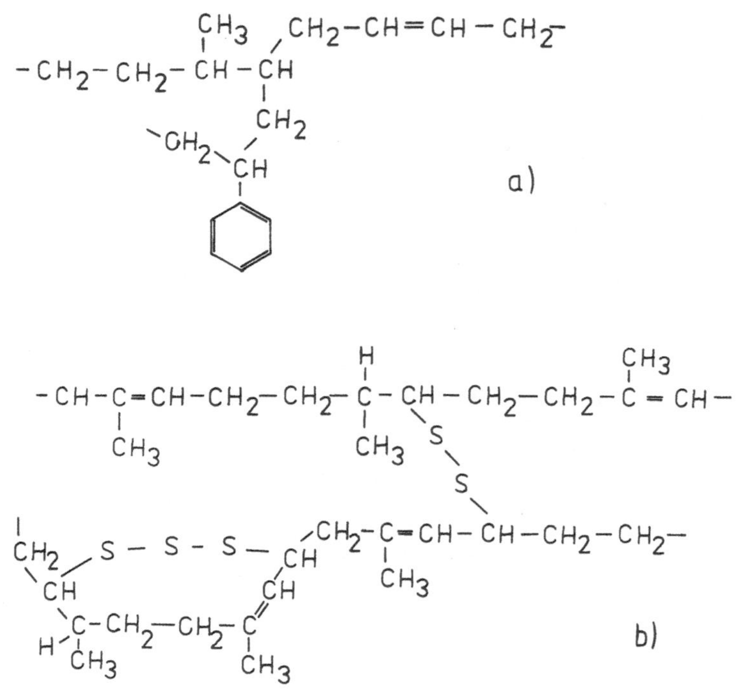

Fig. 1. Part of the proposed structure of a) plasma polymerized polyethylene [1], and b) vulcanized rubber

Some plasma polymerized layers seem to be crystalline. We have some experience with crystalline layers prepared by Klein-Szymanska [3] and Biederman [4, 7] from our institute. These layers are obtained if the plasma polymerization method is used gently. This "gentle" use of the method means approximately that Yasuda's [5] parameter

$$p = W/(F \cdot M) \qquad (1)$$

is low. In Eq. (1) W is discharged power, F is flow rate, and M is molecular mass of the monomer gas used in the experiment.

The structure of ordinary crystalline polymers is very complicated. Areas with well-defined crystal structure and those with nearly amorphous character alternate. The order of regularity differs in different parts of the polymer volume. The order codes of the crystal parts may be different in different crystal blocks, or in different samples according to the conditions of growth [6]. The dominant crystal order code (morphological unit) is the lamella (see Fig. 2). The height d of the lamella depends on the polymer type, and its order of magnitude is 10 nm. The lamellae form larger morphological units, mostly spherulites or fibrillae.

The picture of the polyvinylcarbazole (PVCa) layer prepared by Klein-Szymanska is given in

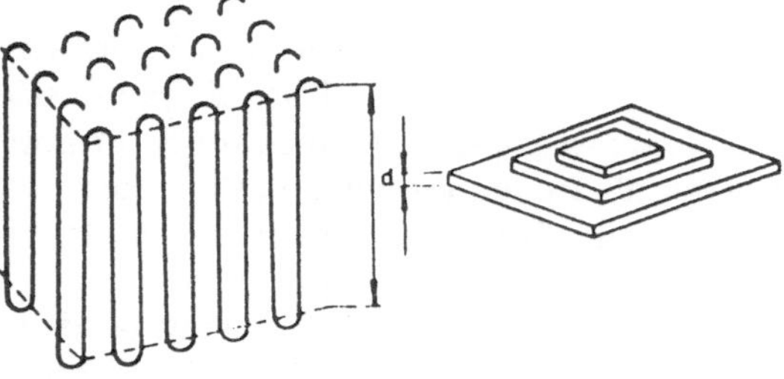

Fig. 2. Schema of a chain folded lamella

Fig. 3. Part a) of the figure may be interpreted as short fibrillae, and in the more detailed picture 3b) the grey creases may be interpreted as lamellae. This interpretation is in accordance with that given by Wunderlich [6] for very similar electron micrographs of classical polymers.

Some examples of the layers prepared by Biederman [7] are given in Figs. 4, 5, 6. These gently prepared layers are relatively thick. The electron micrographs of the layers show complicated structures with some regularity. Such structures are in ordinary polymers interpreted as crystalline. Morphologies visualized in Figs. 4, 5, 6 may be found also in ordinary crystalline polymers [6], therefore,

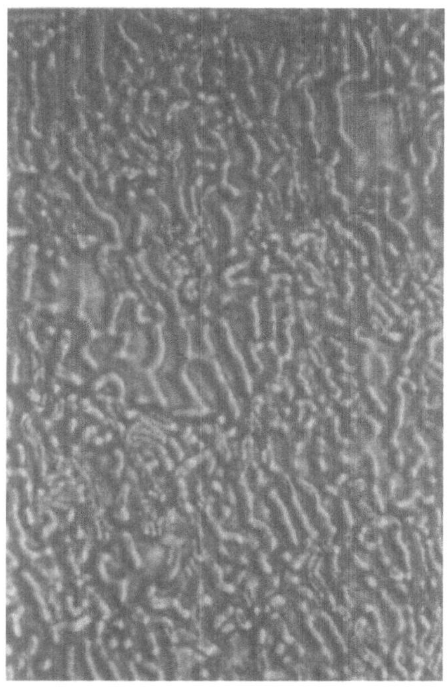

a

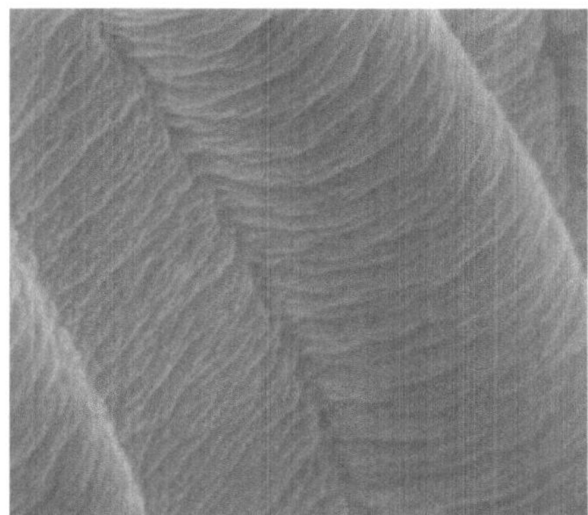

b

Fig. 3. Plasma polymerized PVCa layer. a) Optical micrograph of fibrillar structure (crossed nicols, magnified 90×); b) Scanning electron micrograph (magnified 10000×)

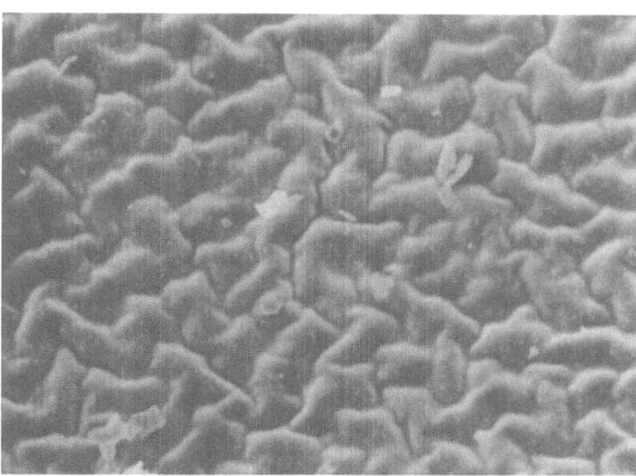

Fig. 4. Scanning electron micrograph (SEM) of the PVCa layer; fibrillar structure (magnified 700×) [4]

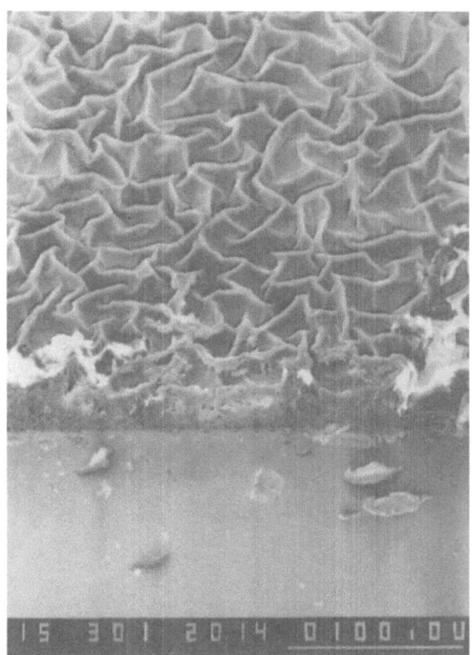

Fig. 5. SEM of the PVCa layer; on the fracture line the thickness of the layer is visible (magnified 350×) [4]

it seems to us that there is no principal difference between ordinary and plasma polymerized crystalline polymers.

The direct mechanical measurement of the surface profile is used in our laboratory to complete the information obtained by optical and electron micrographs. The Surfometer SF 200 of the Planar Industrial Company is used for this purpose. The saphire ball of very small diameter copies the sur-

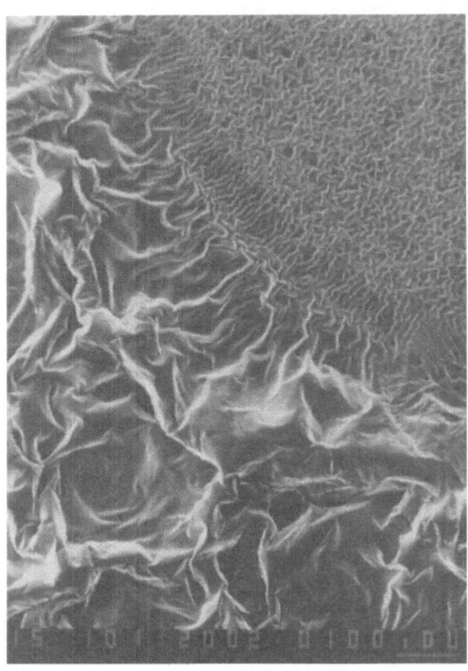

Fig. 6. SEM of the PVCa layer; conditions of growth depend on the substrate (magnified 120×) [4]

face profile along one direction and the result is monitored after magnification on the registration paper (see Fig. 7). The surface graphs indicate large thickness differences (see Fig. 7a) in the case of layers similar to that of Fig. 4, 5, 6. A very regular surface profile was indicated in some parts of the monitored line in the case of polyvinylalcohol (PVAc) layers (see Fig. 7b). The regularity and the order of magnitude of the obtained steps (80 nm) led us to the assumption that lamellar structure was detected. The purity of the monomeric alcohol used may be the reason for the very regular surface structure. This regularity can be seen if the direction of profile monitoring goes along the fibrila. Unfortunately, it was impossible to obtain good preparations for electron micrography from the PVAc layers.

The mean values of the linear deviation Δy from the average thickness of the layer is used as the measure of the layers' roughness

$$R_a = \frac{1}{l} \int_0^l | \Delta y | \, dx \qquad (2)$$

in the surfometric measurements. The distance x is measured along the direction of monitoring.

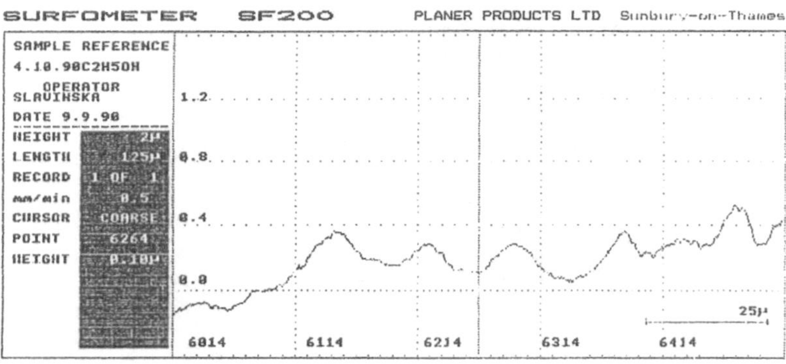

a

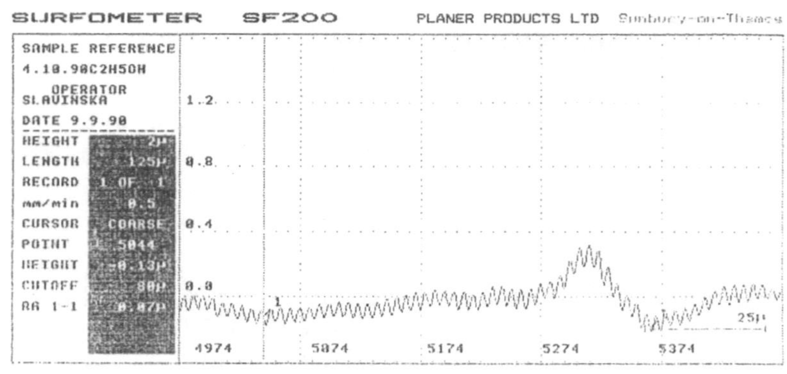

b

Fig. 7. Surface profile of the PVCa layer. The line of monitoring goes a) across the fibrillae, and b) along a fibrilla

The roughness R_a of the plasma polymerized layers depends on the parameter p (see Eq. (1)); in the PVCa layers, we have obtained values between 0.02 μm and 0.1 μm. The roughness of the solution-cast crystalline PVCa layers is only 0.01 μm, and that of the amorphous layers is under 0.01 μm, i.e., under the sensitivity of our surfometer.

3. Conclusion

The plasma polymerized layers may be dealt with as polymer materials. The proposed structure of plasma polymerized polyethylene is similar to that of ordinary polymer network. The presence of non-ethylene units in the plasma polymerized polyethylene networks means that it is not pure polyethylene, which does not discount it as a polymeric material. The presence of irregular units is probable even in the classical polymer networks, only the degree of irregularity is usually lower in these networks than in the plasma polymerized networks.

The structure of crystalline polymers is very complicated and their morphology depends substantially on the preparation conditions. This is valid for both of the discussed materials: the classical polymers and the plasma polymerized layers. All the morphology profiles which have been detected (electron or optical micrographs, surfometry) for plasma polymerized layers have their analogy in the morphology of classical crystalline polymers. Moreover, the very simple test confirms the polymer structure of the plasma polymerized layers — the layers are insoluble.

References

1. Tibitt JM, Bell AT, Shen M (1977) J Macromol Sci Chem 11:139
2. Krakovský I, Havránek A, Ilavský M, Dušek K (1988) Colloid Polym Sci 266:324
3. Nešpůrek S, Slavínská D, Klein-Szymanská B (1979) Eur Polym J 15:965
4. Biederman H, Personal communication
5. Yasuda H (1985) Plasma Polymerization. Academic Press, Orlando
6. Wunderlich B (1973) Macromolecular Physics. Academic Press, New York and London
7. Biederman H, Chudáček I, Slavínská D, Štulík P (1989) Proc ISPC-9, Pugnoiozo, Italy, p 1230

Authors' address:

Dr. A. Havránek
Department of Polymer Physics
MFF UK
V Holešovičkách 2
18000 Praha 8, Czechoslovakia

Discussion

WEICHART:

What did you mean by gentle plasma polymerization conditions and which frequency do you use? Further, is it really possible to deduce the structure of plasma polymerized films from surface topology in your case?

HAVRÁNEK:

Gentle plasma polyermization conditions mean that Yasuda's parameter $P = W/(FM)$ is low and, especially, the power input has been limited to the lowest possible value (some watts). The proper adjustment of the inert atmospheric pressure is necessary, and the frequency of 40 kHz was used.

We do not think that the structure of the layers may by definitely solved from the surface topology, but some interesting results may be compared with similar results obtained for ordinary polymers. In crystalline polymers such morphological studies often serve as a basis for structural interpretation.

Molecular dynamics in ferroelectric liquid crystals: From low molar to polymeric and elastomeric systems

F. Kremer+), S. U. Vallerien+), H. Kapitza*), H. Poths*), and R. Zentel**)

+) Max-Planck-Institut für Polymerenforschung, Mainz, FRG
*) Institut für Organische Chemie der Universität Mainz, Mainz, FRG
**) Institut für Organische Chemie der Universität Düsseldorf, Düsseldorf, FRG

Broadband dielectric spectroscopy (10^{-2} Hz to 10^{10} Hz) is a versatile tool to study the dynamics in ferroelectric liquid crystals: In the frequency regime $<10^6$ Hz the ferroelectric modes, soft- and Goldstone-mode give rise to huge dielectric losses: The softmode is assigned to the fluctuation of the amplitude of the helical superstructure formed by the smectic layers, while the Goldstone-mode corresponds to fluctuations of its phase. In the frequency regime between 10^6 Hz to 10^{10} Hz only *one* relaxation process is observed, the β-relaxation. It is assigned to the hindered rotation of the mesogens around their long molecular axis. At the phase transition from the non-ferroelectric to the ferroelectric state it does *not* split or broaden [1, 2] and its dielectric strength does not decrease as proposed by different theoretical concepts, which are based on the existence of a "free" rotation inside the Sm-A-phase and its strong hindrance in the ferroelectric Sm-C* phase.

For polymeric and elastomeric ferroelectric liquid crystals similar results are observed [3]. Thus it is possible to combine the mechanical properties of polymeric systems with the ferroelectric characteristics and to envisage a completely new field of possible applications (piezoelectric components, pressure sensors, etc.) [4].

References

1. Vallerien SU, Kremer F, Geelhaar T, Wächtler A (1990) Phys Rev A 42:2482
2. Kremer F, Vallerien SU, Kapitza H, Zentel R, Fischer EW (1990) Phys Rev A 42:3667
3. Vallerien SU, Zentel R, Kremer F, Kapitza H, Fischer EW (1989) Macromol Chem 10:333
4. Vallerien SU, Kremer F, Fischer EW, Kapitza H, Zentel R, Poths H (1990) Makromol Chem Rap Commun 11:599

Authors' address:

F. Kremer
Max-Planck-Institut für Polymerenforschung
Postfach 3148
6500 Mainz, FRG

Discussion

GESCHKE:

The piezo-effect obtained on ferroelectric polymer systems was given in values of voltage (some mV). Did you compare the effects quantitatively with those found on PVDF for example?

KREMER:

The observed piezoelectric effect cannot be analyzed quantitatively at the moment. For that it would be necessary to employ a force sensor, which was not available in our studies mentioned above. Furthermore, the sample preparation has to be improved. Up till now the crosslinking reaction was induced by heat in the isotropic phase. So the orientation of the mesogenic groups with respect to the external force is not known at the moment. Thus, a quantitative comparison with the piezocoefficient in PVDF is not possible yet.

SWORAKOWSKI:

You showed that application of a relatively low external field suppresses the Goldstone mode. Does the system recover after the field has been moved, and how rapid is the progress?

KREMER:

In low molar ferroelectric liquid crystals the Goldstone-mode recovers immediately (<1 ms) after the external DC-field is switched off. In polymeric systems this recovery time can last for some days.

KILIAN:

Could you comment about the mechanism of "thermalization" so as to explain the extremely large ε'' values?

KREMER:

The extremely large values of ε'' are caused by the fact that the underlying processes (soft- and Goldstone-mode) are collective on a macroscopic scale (~ 1 µm). But, still, the absolute value of the dissipation of energy is so small that a thermal heating in the sample can be neglected.

Optical properties of LC polymeric films with helical structure

Ya. S. Freidzon and V. P. Shibaev

Chemistry Department, Moscow State University, Moscow, 119899, USSR

Approach to the synthesis, as well as results of phase state study, peculiarities of structure and optical properties of side-chain LC polymers with the helical organization (cholesterics and chiral smectics) and their blends with low molar mass liquid crystals are presented.

Acrylic cholesterol-containing homopolymers and different copolymers containing the chiral (mesogenic and nonmesogenic) and the nematogenic (phenyl benzoate and cyanobiphenyl) groups were studied. The different derivatives of cholesterol and p-alkoxybenzoates were used as the low molar mass liquid crystals. Some of cholesterol-containing homopolymers form the S_A phase within the wide interval from room temperture up to about 155 °C, the cholesteric phase — within interval of 4—5 °C and within the narrow temperature range (0.45 °C) — the blue phase. Structure and optical properties of copolymers as well as the blends of LC homo- and copolymers with low molar mass liquid crystals depend on their compositions. The possibility to form either "monochromic" or "enantiochromic" cholesteric phase is shown. The cholesteric polymeric films are characterized by selective reflection of circular polarized light. The wavelength of selective reflection of light depends on the chemical structure of the chiral and the nematogenic components as well as on their content in copolymers and in blends. The helical twisting power of the chiral component as well as temperature dependence of the cholesteric helix pitch are discussed. The influence of the electric field on the helix pitch is demonstrated.

Authors' address:

Ya. S. Freidzon
Moscow State University
Chemistry Department
Moscow 119899, USSR

Optical and electro-optical properties of polymeric liquid crystals

J. H. Wendorff

Deutsches Kunststoff-Institut Darmstadt

Low molar mass and polymeric liquid crystals display extremely large electro-optical and nonlinear optical responses, due to a combination of orientational long range order and fluidity. These reponses are controlled on the one hand by the intrinsic anisotropic dielectric and optical properties of the phase and on the other hand by their peculiar curvature elastic and viscous properties.

Polymer liquid crystals — although having the disadvantage of longer response times — offer the advantage that the liquid crystalline state and any modulation imposed on this state by external fields, can be frozen in the glassy state.

Polymeric liquid crystals can thus be used in various applications related to the control of light propagation, optical information storage and the manufacturing of holographic optical elements.

Author's address:

J. H. Wendorff
Deutsches Kunststoff-Institut
Schloßgartenstr. 6R
W-6100 Darmstadt

Discussion

FREIDZON:
What can you say about mechanical properties of these polymers? Is it possible to use them without substrate?

WENDORFF:
The mechanical properties of such side chain liquid crystalline polymers are poor. They are usually short-chain molecules, they are not really entangled.

KREMER:
How did you manage to decrease the "writing time" of your system from $\tau = 10$ s to $\tau = 10^{-3}$ s?

WENDORFF:
The writing time is decreased in our systems since the basic process involved is not a collective (as in the case of normal electro-optical switching), but a local one. Furthermore, the intensity/writing time-dependence does not show a threshold behavior, which again leads to a speed-up.

WEICHART:
Do you intend to combine LC-films with "conventional" integrated optics, and how could this possibly work?

WENDORFF:
We intend to do so. We already prepare thin films (1 µm and below) on substrates such as wafers, quarz, glass, etc. We can then modify the local optical properties by laser-light irradiation. This can be done in a very specific way by using a laser scanner.

KILIAN:
Can you obtain, by modified applications of your methods, knowledge about properties of a glass?

WENDORFF:
We know that the induction of cis molecules reduces T_g. We also know that the thermal cis-trans back relaxation displays WLF-behavior. Furthermore, we can sample the local free volume by looking at the trans-cis, cis-trans transitions as a function of the volume required for such transitions. But, of course, this kind of information is also available from other types of investigations.

Orientational order in LCP layers

D. Geschke and G. Fleischer

Universität Leipzig, Sektion Physik

Abstract: Solid state NMR is used to study the orientational order in thin layers of liquid-crystalline polymers (LCP) after a preceding influence of an orienting magnetic field $\vec{B}$. — Samples were produced of polysiloxanes with mesogenic groups in the side chains by melting the polymer between glass plates coated with a polyimide film. For orientation they were placed in the external magnetic field and slowly cooled down below the glass transition temperature. — Good domain orientation could be obtained if $\vec{B}$ acted parallel to the glass plates. Alignment could be realized only in part if $\vec{B}$ was directed perpendicular to them. There is no remarkable orientation effect of the glass plates themselves.

Key words: Liquid-crystalline polymers; polymer layers; orientational order; solid-state NMR

Introduction

In recent years various papers have been published that deal with nuclear magnetic resonace (NMR) investigations of polymers with mesogenic units in the main and side-chain, respectively; most of them concentrate on bulk material [1—6]. From a practical point of view, thin layers of liquid-crystalline polymers (LCP) are of special interest, because they can be used, for example, as optical storage disks, polymer electrets, and polymer photoconductors.

All applications are connected with special properties of the materials such as high transmission in the visible range, high optical and electrical homogeneity, and mechanical rigidity and tight thickness tolerance, etc.

The liquid-crystal state is intermediate to crystal and liquid phase and displays molecular organizations in which orientational ordering is the dominant theme, while the long-range positional ordering characteristic of a crystal is absent. It is orientational order which is successfully exploited in LCP to give, for example, electro-optical effects for information storage or piezo- and pyroelectric effects for sensor application.

In many ways, in the study of anisotropy within noncrystalline polymer systems the greatest problem generally is its definition. For example, what is meant by molecular orientation in an LCD in which just the side groups are mesogenic? This might imply the anisotropy in the trajectory of the molecular backbone, or it could mean the alignment of the side-chain alone.

There is a wide range of appropriate experimental techniques available to probe orientation, but our emphasis will be an exploitation of NMR procedures.

The evaluation of molecular orientation in LCP is, of course, only one step in the process of understanding the origin of specific macroscopic properties.

Polymers which have been subjected to external fields, whether surface stress, magnetic, or electric may show a preferred alignment of structural units either about, or normal to the applied field direction. The quality of alignment achieved is generally not perfect in relation to the field axis.

Theory

When the probability distribution of the orientation of units is symmetrical about any axis, then it can be referred to that axis without the necessity of an external reference frame. Such an axis is known as the director. Here, we shall restrict ourselves initially to situations where this grouping occurs symmetrically about only one axis, in other words, to uniaxial systems.

Figure 1a shows the orientation of a director $\vec{d}$ within the laboratory frame x, y, z.

Uniaxiality should be equal to "transverse isotropy", i.e.,

— within a domain characterized by the director $\vec{d}$ the (molecular) axes $\vec{a}$ are uniformly distributed on a cone around the director axis $\vec{d}$

$$(\cos\Theta_a = \vec{a},\vec{d}) \quad \vec{a},\ \vec{d}\text{: unit vectors}$$

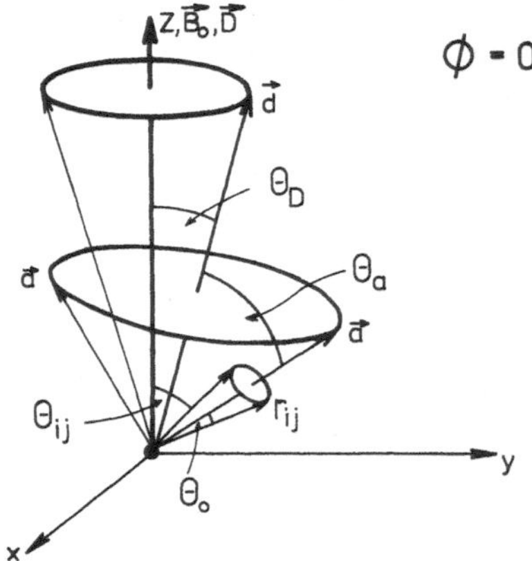

Fig. 1a. Orientation of director $\vec{d}$ within the laboratory frame x, y, z.
θ, φ: Euler angles;
Φ: angle between $\vec{B}_0$ and $\vec{D}$ ($\vec{B}_0$, z-plane);
Θ: angle between $\vec{B}_0$ and $\vec{d}$.

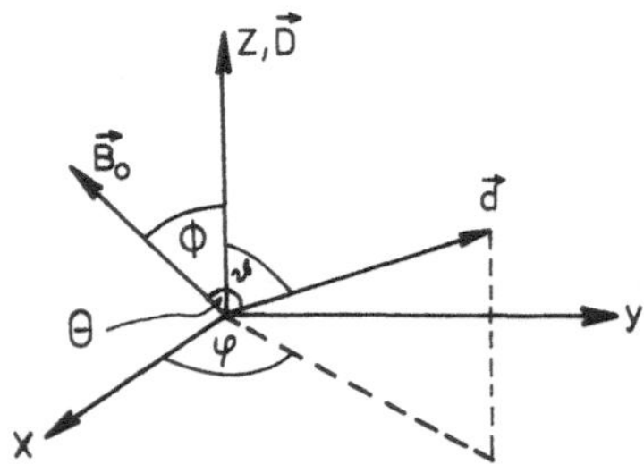

Fig. 1b. Position of $\vec{D}$, $\vec{d}$, $\vec{a}$ and $\vec{r}_{ij}$ within the frame x, y, z.
$\vec{r}_{ij}$: internuclear vector;
$\vec{a}$: side-chain axis;
$\vec{d}$: director axis;
$\vec{D}$: preferred director axis

— the director distribution is axially symmetric around a preferred axis $\vec{D}$

$$(\cos\Theta_D = \vec{d},\vec{D}) \quad \vec{D}\text{: unit vector ,}$$

which can be described, e.g., in a very simple way [7]:

$$P(\theta) = (1-U)\,\frac{\delta(\theta)}{\sin\theta} + \frac{U}{2}\ ; \tag{1}$$

U denotes that part of the sample in which the $\vec{d}$-vectors are distributed isotropically . $\delta(\theta)$ means Dirac's δ-function.

For the simplest case of a two-proton sytem the resonance spectrum is a doublet according to the two orientations of a proton dipole with respect to the static magnetic field $\vec{B}_0$. Line splitting is given by the well-known equation

$$\Delta\nu_{ij} = \frac{3}{4}\,\frac{\gamma^2\hbar}{r_{ij}^3}\,\frac{\mu_0}{4\pi}\,(3\cos^2\Theta_{ij} - 1)\ , \tag{2}$$

where γ is the gyromagnetic ratio of protons; $\vec{r}_{ij}$ is the internuclear vector; $\hbar$ is Planck's constant divided by 2π; Θ_{ij} is the angle between $\vec{r}_{ij}$ and $\vec{B}_0$, and μ_0 is $4\pi \cdot 10^{-7}$ Vs/(Am).

However, in the real case an ensemble of two-spin-systems undergoing molecular motion has to be considered, and instead of Eq. (2), one has to calculate

$$\Delta\nu = \Delta\nu_0\langle(3\cos^2\Theta_{ij} - 1)\rangle\ , \tag{3}$$

with

$$\Delta\nu_0 = \frac{3}{4}\,\frac{\gamma^2\hbar}{\langle r_{ij}^3\rangle}\,\frac{\mu_0}{4\pi}\ . \tag{4}$$

Assuming, in addition to uniaxiality of axes $\vec{a}$ with respect to $\vec{d}$ and $\vec{d}$ with respect to $\vec{D}$ (see above), fast reorientation of side-chain axes (i.e., no preferred direction of two-spin-systems perpendicular to $\vec{a}$), one obtains from Eq. (3), using the relation

$$\cos\theta = \cos\theta\cos\varphi + \sin\theta\sin\varphi\cos\varphi\ , \tag{5}$$

and known transformation rules

$$\Delta\nu = \Delta\nu_0\left\langle\frac{1}{2}\,(3\cos^2\Theta_0 - 1)\right\rangle SS'(3\cos^2\varphi - 1), \tag{6}$$

with the order parameter S

$$S = \frac{1}{2} \langle(3\cos^2\Theta_a - 1)\rangle , \quad (7)$$

and the parameter of orientational order S':

$$S' = \frac{1}{2} \langle(3\cos^2\Theta_D - 1)\rangle$$

$$= \frac{1}{2} \int_0^\pi (3\cos^2\theta - 1)p(\theta)\sin\theta d\theta . \quad (8)$$

The latter describes the alignment of domains by the external magnetic field. Full alignment of directors along $\vec{D}$ ($U = 0$) corresponds to $S' = 1$.

Experimental part

Polymers of the following type were under consideration

$$\begin{array}{c} + \\ \mathrm{O} \\ | \\ \mathrm{CH_3{-}Si{-}(CH_2)_n{-}X{-}R} . \\ +_{35} \end{array}$$

They were synthesized by B. Krücke (Martin-Luther-Universität Halle).

The NMR measurements were performed at a proton resonance frequency of 60 MHz using a commercial spectrometer. The electronic assembly is characterized by a dead time of 6 ... 7 μs. In order to guarantee a good signal-to-noise ratio, at least 500 signals were accumulated in each case.

All samples were proved for molecular alignment using a polarizing microscope.

Results and discussion

In Fig. 2 typical spectra are shown, measured in dependence upon the angle Φ between the normal $\vec{n}$ of the film and the static NMR field $\vec{B}_0$ (cf. [8]).

In case A it was tried to orient the domains along the plane of the glass plates, and in case B, perpendicular to them. For comparison, a spectrum is given for a sample which was cooled from the isotropic state without influence of a magnetic field (case C).

Considering the order of the line-splitting indicated, one can conclude that the main part of the spectra is caused by the nuclei of the mesogenic units, whereas the central component of the signals is interpreted to be due to the more or less mobile alkyl groups of the side-chains. However, some hints are given in literature [9] that protons of non-oriented side-chains also contribute to the central peak.

Polymer no.	n	X	R	Phase transition temperatures in °C
1	3	O-⬡-COO-⬡	O-C_5H_{11}	g43S_B52S_A177i
2	3	O-⬡-COO-⬡	O-C_6H_{13}	g62S_A174i
3	4	COO-⬡-⬡	CN	g14S_A132i

(g: glassy, s: smectic, i: isotropic phase)

Samples were produced by melting the polymer between 4 mm × 20 mm glass plates coated with a thin polyimide film (samples 1,2). In connection with test experiments with electrical fields a metallization of the glass plates was necessary. The thickness of the polymer layer was approx. 50 μm.

For orientation, the samples were placed in a magnetic field $\vec{B}$ of 2.1 T and slowly cooled from a temperature of T_i + 20 K below the glass transition temperature T_g (rate: 3 K/h).

Figure 3 gives a series of spectra obtained with sample 3.

Because of the bad resolution in most cases, we used the width at 50% of the maximum amplitude values as a parameter. The corresponding angular dependencies are shown in Fig. 4. $\Delta\nu$-values, which could be estimated directly, are included and have been used for calibration.

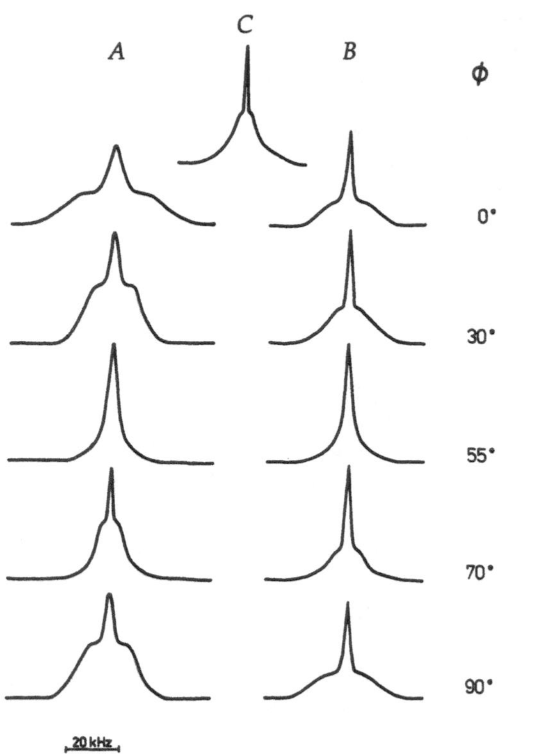

Fig. 2. Experimental spectra of sample 1 oriented in the magnetic field $\vec{B}$.
A: $\vec{B}$ parallel to the polymer layer;
B: $\vec{B}$ perpendicular to the polymer layer;
C: Unoriented sample (sample 2)

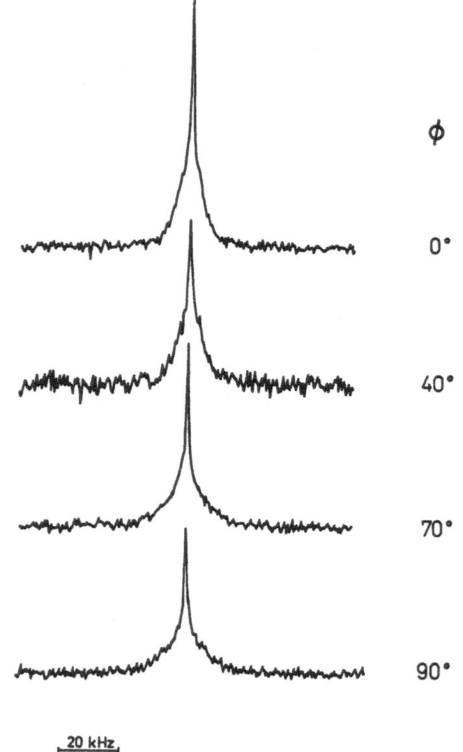

Fig. 3. Experimental spectra of sample 3

Using known lattice parameters from the literature (e.g., [10]) one obtains for the ortho-proton pairs $\langle r_{ij}\rangle$ between 0.224 nm and 0.239 nm and, therefore, $\Delta\nu_0 \approx 25.4$ kHz. Further, we admitted a small angle between the side-chain axis $\vec{a}$ and the para-axis of the benzene rings; it should range somewhere between 5° and 10° [10], leading to $\langle 1/2(3\cos^2\Theta_0 - 1)\rangle \approx 0.955$.

In case *A* the alignment of the mesogenic groups is nearly perfect, as could be detected with the polarizing microscope. In accordance with this optical result, we assume $S' \approx 1$, allowing us to estimate $S \approx 0.66$; such a value is typical for smectic systems. Taking into account the experimental errors and considering the average values of $\langle r_{ij}\rangle$ and Θ_0 used above, *S* is of the same order of magnitude as in bulk material.

It is interesting to note that the line-widths measured at 50% of the maximum amplitude value also reflect the theoretical angular dependence with a minimum value at $\Phi \approx 55°$ (cf. Fig. 3). The spectra in case *B* (Fig. 2) are significantly different and more similar to those of the isotropic case. Assuming the same *S*-values (this may not be quite correct), it is possible to estimate the parameter of orientational order to be $S' \approx 0.50$. This means that approximately 50% of the domains could not be aligned by the magnetic field. Unfortunately, this result could not be proved directly by optical methods as in case *A*.

Figure 5 shows the corresponding angular dependence for sample 3.

As can be seen, there is only a weak indication of orientation of domains, but it is necessary to note that internal fields produced by the dipoles compete with external ones and suppress side-chain alignment by the magnetic field. The electric field of a CN-dipole at a distance $r = 2.5$ Å can be estimated to be $\vec{E} = 5 \cdot 10^7$ V/cm; it is much higher than the electrical fields we used in test experiments. Further investigations are in preparation.

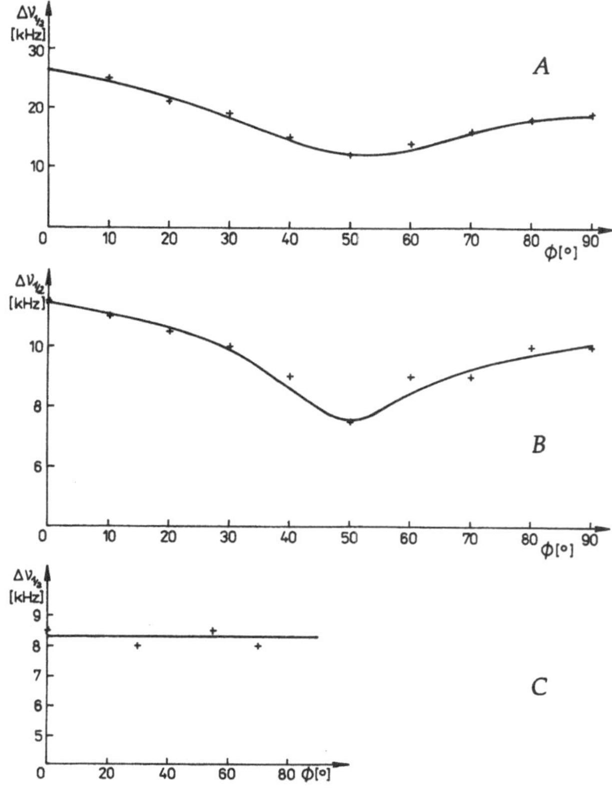

Fig. 4. Angular dependence of $\Delta\nu_{1/2}$ of samples 1 and 2.
A: $\vec{B}$ parallel to the polymer layer;
B: $\vec{B}$ perpendicular to the polymer layer;
C: Unoriented sample (sample 2)

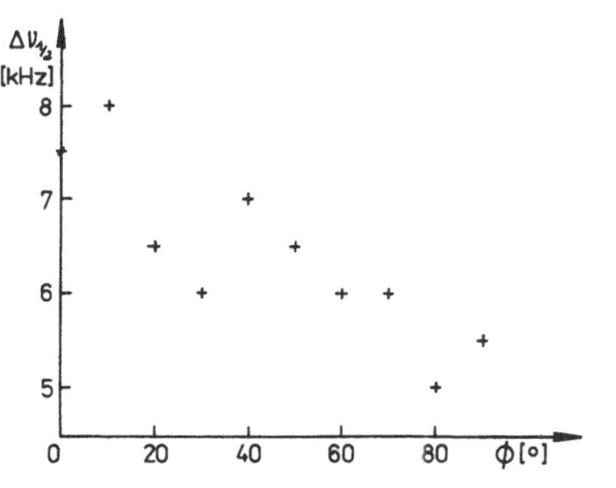

Fig. 5. Angular dependence of $\Delta\nu_{1/2}$ of sample 3

Conclusions

The following conclusions can be drawn:

— there is no remarkable (orientation) effect of the polyimide coated glass plates;
— using an external magnetic field of 2.1 *T*, good domain orientation can be obtained if $\vec{B}$ acts parallel to the glass plates, however, internal fields caused by electric dipoles remarkably suppress alignment; and
— alignment can be realized only in part if $\vec{B}$ acts parallel to the normal $\vec{n}$ of the plates.

References

1. Piskunov MV, Kostromin SG, Stroganov LB, Shibaev VP, Plate NA (1982) Makromol Chem Rapid Commun 3:443—447
2. Roth H, Krücke B (1986) Makromol Chem 187:2655—2662
3. Boeffel C, Spiess HW, Hisgen BU, Ringsdorf H, Ohm H, Kirste RG (1986) Makromol Chem Rapid Commun 7:777—783
4. Pschorn H, Spiess HW, Hisgen B, Ringsdorf H (1986) Makromol Chem 187:2711—2723
5. Perry BC, Koenig JL (1988) Polym Prepr (Am Chem Soc, Div Polym Chem) 29:50—51
6. Kireev EV, Stroganov LB, Gubina T, Kostromin SG, Talroze RV, Shibaev VP, Plate NA (1989) Vysokomol Soed B31:261—263
7. Frieser A, Hillner B, Schmiedel H (1981) Wiss Z Karl-Marx-Univ Leipzig, Math-Naturwiss Reihe 30:156—163
8. Geschke D, Fleischer G, Dressler R (1989) Makromol Chem, Rapid Commun 10:595—599
9. Geschke D, Hempel G (1989) Proceedings of the "9. Tagung Polymerphysik", Potsdam 24.—28.4., Phys Gesellschaft der DDR, Berlin p 14
10. Limmer S, Schmiedel H, Hillner B, Lösche A, Grande S (1980) J Phys (Les Ulis, Fr) 41:869—878

Received January 11, 1991
accepted February 4, 1991

Authors' address:

Doz. Dr. D. Geschke
Universität Leipzig
Sektion Physik
Linnestraße 5
O-7010 Leipzig, FRG

Discussion

WENDORFF:

Can you also use this experimental method to follow the kinetics of texture formation?

GESCHKE:

Up to now we measured the angular dependence of the NMR parameters mentioned above at temperatures for which rigid lattice — like conditions could be assumed. But similar experiments can be done if the kinetics of texture formation, i.e., the alignment of domains under the influence of the magnetic field, is of special interest. However, in such cases additional motional processes must be considered which also affect line-splitting and line widths.

KILIAN:

Could you draw conclusions from the line-shape analyses about the distribution functions?

GESCHKE:

In the case of thin layers of liquid-crystalline materials, we estimated only the second-order Legendre polynomials S and S' in order to characterize the distribution functions. This was because of the relatively poor signal-to-noise ratio.

For a more detailed study, higher experimental accuracy is required.

Thin polymeric layers for spatial light modulators

R. Gerhard-Multhaupt

Heinrich-Hertz-Institut für Nachrichtentechnik Berlin GmbH, Berlin, FRG

Abstract: The use of thin polymeric control layers in spatial light modulators for optical applications is surveyed. Three major groups of light modulators with polymers as electro-optic components are introduced. Light control by diffraction is the principle of deformable light modulators, whose main varieties are membrane light modulators, thermoplast films, and elastomer layers; considerable research and development efforts, mainly during the last two decades, led to a wide range of proposed devices and a good understanding of their properties. The more recently invented polymer-dispersed liquid crystals operate with scattering as light-control mechanism; even though these materials are already found in practical applications, their investigation and development is far from complete. Finally, polymer films with electro-optic effects that are caused by molecules or molecular groups with large second- or third-order dielectric susceptibilities are briefly discussed; in such materials, light control is achieved by polarisation-dependent refraction or birefringence.

Key words: Spatial light modulator; polymer membrane; thermoplast; elastomer; polymer-dispersed liquid crystal; electro-optic layer

1. Introduction

Spatial light modulators [1, 2] are key components of optical systems for light-valve image projection, reconfigurable optical interconnection, optical information processing, image conversion, etc. As schematically depicted in Fig. 1, light modulation is usually achieved by means of electrically controlled diffraction, scattering, or birefringence. Electronic or optical addressing with passive or active matrices or with photoconducting elements, respectively, provide the required electric fields. The application of polymer layers in such two-dimensional spatial light modulators offers several advantages such as a wide variety of available effects and materials, great flexibility with respect to different geometry and performance requirements, many possibilities for optimisation by means of materials engineering, and relatively easy preparation in most cases. In the following survey, three basic types of spatial light modulators with polymer-based light-control layers are illustrated by considering devices described in the literature.

2. Deformable light modulators: Light control by diffraction

Deformable light modulators vary the phase distribution of the read-out light. The resulting diffraction is evaluated by a schlieren-optical system or by suitable apertures. Commercial light-valve applications such as the well-known Eidophor [3] and Talaria [4] large-screen projectors are based on this principle. In these devices, light modulation is achieved by a transparent fluid layer whose surface deformation is controlled by electron-beam-deposited electric charges. While the fluids contain polymer additives for stabilisation, truly polymeric control layers are found in membrane, thermoplast, and elastomer spatial light modulators.

Mechanism	Diffraction	Scattering	Birefringence
Physical requirement	Systematic optical path variations	Wavelength-sized scatterers	Polarisation-related refractive index
Optical effect	Distribution of minima & maxima	Randomisation of light directions	Change of light polarisation
Schematic depiction			

Fig. 1. Schematic representation and basic features of non-absorptive light-control mechanisms suitable for use with polymers

2.1 *Membrane light modulators*

A membrane light modulator (MLM) for optical-computer applications was already described more than two decades ago [5, 6]. Its membrane typically consists of a 100-nm-thick film made from collodion (pyroxyline nitrocellulose solution) and coated with a mirror electrode. As shown in Fig. 2, the polymer membrane is tightly stretched over a 1-μm-thick silicon-oxide layer with circular perforations that extend down to a glass substrate with chrome-gold electrode strips. Application of, for example, 50 V between membrane mirror and substrate electrode(s) leads to an electrostatic deflection of the membrane into a parabola. The membrane light modulator was employed in a coherent optical processing system capable of performing fast Fourier transforms [5, 6].

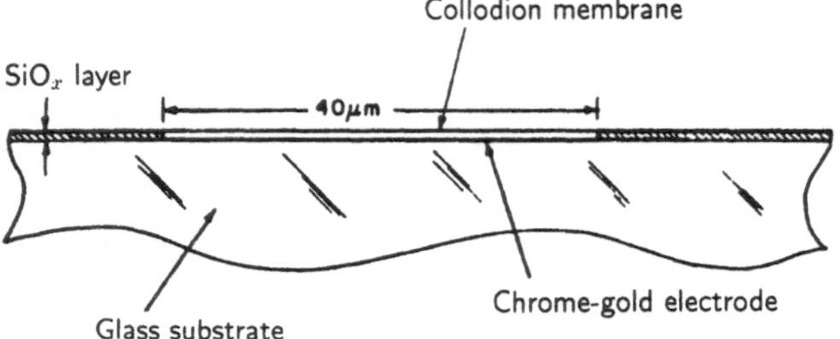

Fig. 2. Cross-section of the membrane light modulator with electrical addressing by means of electrode strips [5]

While the electrical control of the membrane light modulator was attained with passive electrode strips, the development of silicon-based microelectronics made it possible to use active transistor matrices for the electrical addressing of spatial light modulators. The combination of silicon semiconductor technology with metallised nitrocellulose membranes led to the so-called deformable mirror device (DMD) [7, 8]. In this device, the spacers may be fabricated by means of etching processes from polysilicon or dielectric layers that are required for matrix operation anyway. An alternative version of the deformable mirror device incorporates optical addressing of the silicon chip through the partially transparent membrane electrode [9]. More recently, the polymer membrane was replaced [10] by cantilever beams that are manufactured with anisotropic etching techniques now frequently utilised in silicon micromechanics; the name "deformable mirror device" was transferred to the new version [10].

Intended applications of the old and new DMDs include light-valve projection displays, printing heads, and optical processing systems for correlation, switching, and neural-network implementation [7—10]. A third variant of the same basic principle is the optically addressed photoemitter membrane light modulator (PEMLM) [11, 12] for fast optical information processing [13]. Its indium-coated nitrocellulose membrane is stretched over a microchannel plate, which amplifies and guides the electron current that is emitted from the illuminated

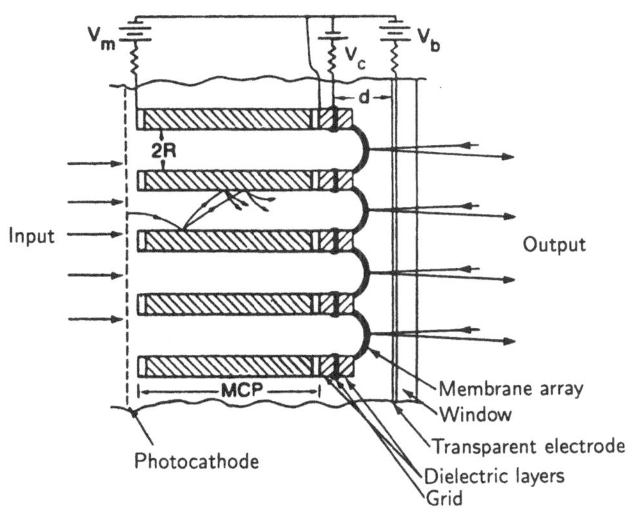

Fig. 3. Cross-section of the optically addressed photoemitter membrane light modulator [11—13]

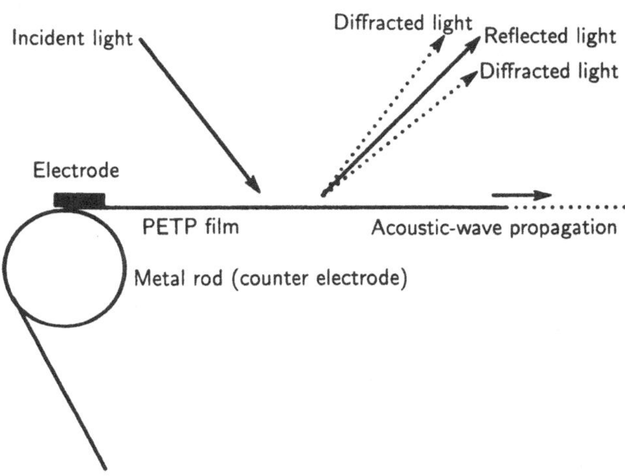

Fig. 4. Schematic arrangement of the thin-film travelling-wave light modulator [17]

areas of a photocathode. Thus, as depicted in Fig. 3, the nonconducting membrane is charged and pulled toward the transparent electrode of the read-out window by electrostatic forces [11]. Recently, in view of the non-ideal properties of nitrocellulose, alternative membrane materials were investigated [12]; the resulting optimised version of the photoemitter membrane light modulator contains a parylene membrane on a silicon-oxide (SiO_x) substrate that has been coated with a thin layer of poly[bis(trifluoroethoxy)phosphazene] (PTFEP) in order to lower its surface energy and thus prevent the membrane from sticking to the inside walls of the substrate perforations [14].

Light diffraction by means of acoustic waves propagating in a polymer film is the operating principle of the thin-film travelling-wave light modulator [15—17]. As shown in Fig. 4, a polyethylene terephthalate (PETP) film is stretched between two metal rods, and acoustic waves are generated by applying radio-frequency signals between one of the rods and a small electrode on the opposite surface of the film. Upon propagation, these waves represent a travelling phase grating because of the refractive-index variations brought about by compression and rarefraction in the polymer film and also by local film bending. Modulation of the wave-generating signal causes a corresponding modulation of the travelling phase grating. Diffraction efficiencies as high as 30% were observed for reflected light in the first diffraction order [17].

The device was employed for the determination of relaxation times and hysteresis effects in the polar PETP membrane material [18]. Furthermore, an optical correlator was demonstrated, in which two parallel travelling waves carrying different input modulations are generated by two separate electrodes on the polymer film; a laser spot is diffracted by the first wave according to its input signal and then redirected and spread in order to illuminate the second wave over the required time interval; after the second diffraction, a linear photodetector array integrates the output light, whose intensity is proportional to the convolution of the two input signals [19]. Autocorrelation is also possible if only a single travelling wave is being propagated and if the read-out light is used twice on it, first as a spot and then as a line of light [19].

2.2 *Thermoplast films*

Polymer membranes for light modulation exhibit relatively fast response, but their storage capability is limited by the stability of the electric charges on the deformable film and on the addressing electrode. With thermoplast films, the deformation can be "frozen", which makes such light modulators ideal for the storage of recorded information [20]. Because of the required thermal cycle, the time that is necessary for complete recording lies in the range of several seconds. The softening of the thermoplast material (e.g., polystyrene [21]) leads to relatively high sensitivities: Only moderate electric fields are needed for the deformation of the polymer surface.

Serial writing of the information by means of an electron beam [20] or a moving corona discharge [21] is time-consuming. Much faster recording becomes possible with parallel optical writing. In this case, an additional photoconductor layer effects the light-induced local changes of the electric field. A typical operation cycle of such a two-layer device is schematically shown in Fig. 5 [2]. As a consequence of the large relative deformations possible with molten thermoplast films, the film thickness can be small and the spatial resolution correspondingly high [22]. Not only the deformable thermoplast, but also the addressing photoconductor may be a flexible polymer material (e.g., poly-N-vinylcarbazole with a suitable photosensitive dye) [23] so that the film material may be rolled. Recent optimisation of the thermoplast recording device concerns the dye [23], the charging process [24], and the thermal cycle [23].

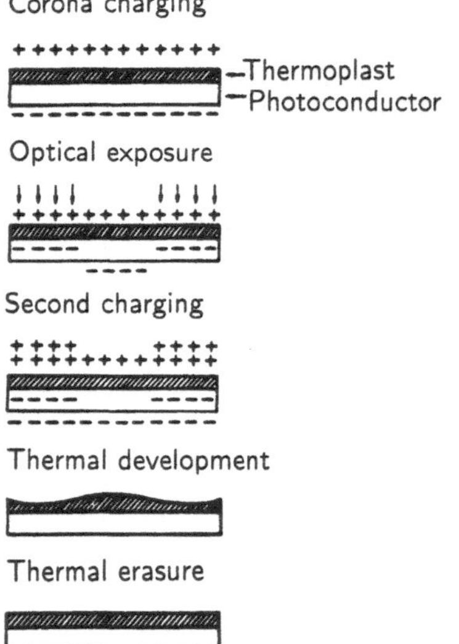

Fig. 5. Typical operation cycle of a photoconductor-addressed thermoplast device for image or hologram recording [2]

Proposed applications of thermoplastic light modulators include the so-called Lumatron [25], a high-resolution storage and projection display with electron-beam addressing, hologram recording [26, 27] for non-destructive testing and other industrial uses, and coherent optical information processing [28] with very high resolution. The area of thermoplast films can be fully employed for light modulation, since no mechanical supports as with membrane light modulators are required. This advantage, together with the very high resolution of approximately 1000 line pairs per millimeter, make thermoplast devices very suitable for the writing and reading of holographic diffraction gratings at moderate speeds [26—28].

2.3 Elastomer layers

Complete use of the full modulator area (as with thermoplast films) and reasonable speed of response (as with polymer membranes) are combined in the case of elastomeric light-control layers [29—54]. In most such devices, the elastomer layer carries a mirror electrode: Deposition of a thin pellicle or membrane before metallisation usually leads to much better optical quality of the thin mirror [50]. In a metallised viscoelastic spatial light modulator, whose operation is sketched in Fig. 6,

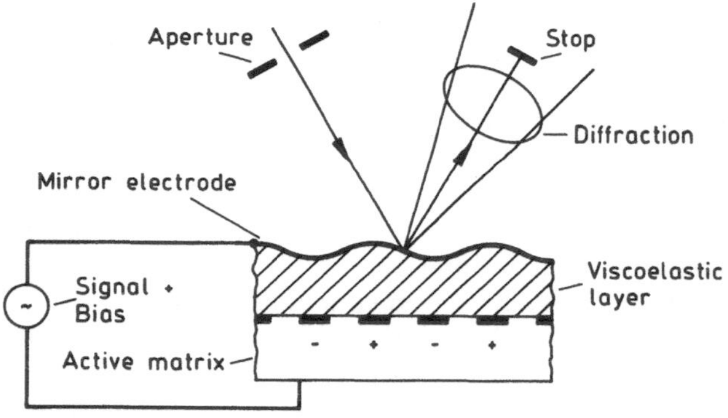

Fig. 6. Principle of light modulation by means of a metallised viscoelastic control layer addressed with an active matrix [51]

the reflective electrode and its underlying pellicle are supported by a soft viscoelastic layer [50]. A periodic or nonperiodic voltage distribution on the addressing device behind the elastomer causes a corresponding surface deformation of the mirror and, thus, also diffraction of the output light; a suitable schlieren-optical imaging system passes only the required diffraction orders and thus transforms the phase modulation generated by the deformed modulator into an intensity modulation [51].

Amplification of the light intensity for image projection was the purpose of a photoconductor-addressed light valve, whose concept followed from research on oil-film light modulators [29, 30]. Polyvinylchloride softened with a plasticiser was proposed as the viscoelastic material to be coated onto a selenium or zinc-sulphide photoconductor [29]. For use in a storage display, the addressing device consisted of an electron-beam-charged mica sheet inside a cathode-ray tube, while the elastomer and mirror-electrode materials were not specified [31, 32]. Around 70 such storage-display tubes were fabricated, and reasonable life times could be achieved [32].

A variety of deformable light modulators based on an elastomer layer were suggested and investigated under the name Ruticon [33—42]. Apart from the most successful gamma-Ruticon with a metallised elastomer [35—37, 39—42], other Ruticon types with surface charging by means of a conductive liquid, a corona or a glow discharge were also studied [33, 34, 38]. Optical addressing of poly-N-vinylcarbazole [35, 37] and amorphous-selenium [40] photoconductors is employed with all Ruticon devices, and the input light illuminates the photoconductor layer either directly or via a fiber-optic faceplate [36]. Extensive theoretical [35, 39, 40, 43] and experimental [33, 35, 37, 39, 40, 42] investigations of the device behaviour led to a detailed understanding of the physics of the Ruticon. Its proposed applications include image projection [34, 36, 41] and optical information processing [38, 43].

Direct electrical addressing of the metallised viscoelastic light-control layer by means of an active matrix forms the basis of light-valve devices for television applications [44—51]. Addition-crosslinked two-component silicone elastomers and collodion or other pellicles are employed in these devices [50], the mirror electrodes typically consist of aluminium or silver. The theoretical modelling of deformable-elastomer light modulators was improved by adding the theories of linear viscoelasticity and plate bending [45—49] to the earlier descriptions that were based on elastic shear and effective surface tension only. Pellicle formation by means of a glow-discharge treatment of the elastomer surface represents an interesting alternative to the collodion pellicle [52]. Total internal reflection at the surface of uncoated elastomer layers may be used instead of the mirror-electrode reflection [53, 54]; more sensitive devices result from this approach, but the required imaging via a prism at a large angle, the unavoidable light exposure of the elastomer, and the slower response may be disadvantages.

3. Polymer-dispersed liquid crystals: Light control by scattering

Electric-field-controlled light scattering is the light-control mechanism of the recently introduced polymer-dispersed (or polymer-encapsulated) liquid-crystal (PDLC) light modulators [55—81]. Apart from being exciting research objects, the complex two-phase PDLC materials, sometimes also called nematic curvilinear aligned phase (NCAP) [59], are already found in a number of practical applications. A display based on electrically controlled index matching between a polymer material and a liquid crystal was first proposed only a decade ago [55], while its basic optical principle had already been conceived in the nineteenth century [82]. During the last seven years, the development of a fabrication technology for liquid-crystalline microdroplets dispersed in a polymer matrix [56] led to a rapid evolution of scientific investigations and practical devices.

The operating principle of PDLC light valves [57] is illustrated in Fig. 7: Without an electric field, the liquid crystal in the microdroplets possesses random orientation, which leads to light scattering in the film; a suitable electric field orients the LC directors perpendicular to the film surface and thus makes the film transparent if liquid crystal and polymer matrix are refractive-index-matched. For the polymer matrix, resins such as a mixture of epichlorohydrin and bisphenol A to be cured with a fatty polyamine agent [57], polycarbonate [63], polyvinylformal [63], epoxy mixtures [63, 68, 74], or polyvinyl alcohol [69, 75] are employed. The liquid-crystal phase typically consists of commercially available LC materials such as 5CB [57, 62], 7CB [62], E7 [62, 63, 68, 69, 74], E8 [62], or guest-host mixtures [69, 79]. Beside chemical curing of the (prepolymer + liquid-crystal) emulsion or solution, thermally induced phase separation [63], photo-initiated polymerisation with ultraviolet light [69, 74, 80] and the recently suggested electron-beam curing [81] are utilised. The size distribution and the number density of the liquid-crystalline micro-

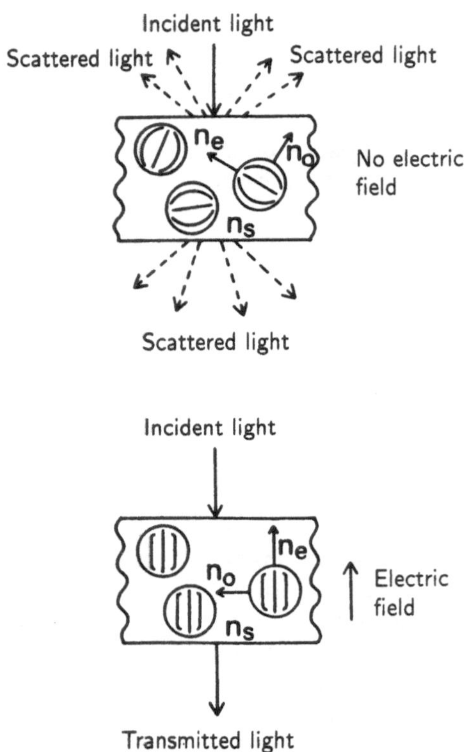

Fig. 7. Light-control mechanism in a polymer-dispersed liquid crystal where the ordinary refractive index n_0 of the liquid crystal is matched to the refractive index n_s of the polymer matrix [57]

droplets are strongly influenced by the polymerisation kinetics [59, 74] and may thus be quite easily varied.

Theoretical investigations of PDLC effects include the modelling of light scattering from a nematic droplet with two different approaches [60, 66], a calculation of light transmission [70], and a study of temperature effects in PDLC films [72]. A considerable body of experimental results was obtained by means of optical measurements [7, 59, 61—63, 65, 69—71, 74, 75, 78], electron microscopy [57, 59, 64, 68, 70, 72, 74], and deuterium [64] and proton [68] nuclear magnetic resonance. Studies of the time behaviour [57, 74, 75] and of the reorientation dynamics inside the microdroplets [69] demonstrated that PDLC light modulators are usually fast enough for video applications.

PDLC films are already commercially available and are being studied for several device applications such as direct-view high-resolution displays [58], large-area light valves [59], full-colour projection with three PDLC light valves [67, 76, 78], frame-sequential colour projection with a single PDLC device [77], and photoconductor-addressed spatial light modulators for optical information processing [73]. In addition to the standard mode, in which the scattering PDLC material is switched to a more or less transparent state, the large Kerr effect in very small LC droplets [74] and optically induced bistability in polymer-dispersed guest-host liquid crystals [79] were proposed as light-modulation mechanisms.

4. Electro-optic polymer layers: Light control by birefringence

During the last decade, the nonlinear optical properties of polymer layers, which contain monomer units, side groups, or guest molecules with large second- or third-order dielectric susceptibilities, became a major focus of attention in polymer science [83—96]. Naturally, such polymers also possess electro-optic properties and may be especially suitable for wave-guide and other thin-film devices [90—96]. So far, the electro-optical performance of polymeric layers does not quite reach that of the best inorganic materials, but their easier preparation and superior mechanical properties are significant advantages in many possible applications. Therefore, the search for better electro-optic polymer materials is continuing.

In the beginning, possibly useful effects and suitable materials were identified, theoretically analysed, and compared [83—85, 87—92]. At the same time, several concepts for potential device applications were proposed [86, 91, 93—95]. Guest-host systems consisting of glassy polymers that are doped with highly optically nonlinear dyes were suggested and prepared [84, 88, 89, 97]. They only exhibit nonlinear behaviour when the random orientation of the dye molecules is transformed into a preferential orientation (usually perpendicular to the film surface) by a symmetry-breaking poling process. This poling procedure, whose main steps are schematically depicted in Fig. 8, is identical to the thermal-poling process of electret films [98]. By heating the polymer to temperatures above its glass-rubber transition and cooling it with the poling field still applied, the preferential orientation of the molecular dipoles can be frozen.

As with other polymer-electret films, corona poling offers advantages over thermal poling: Max-

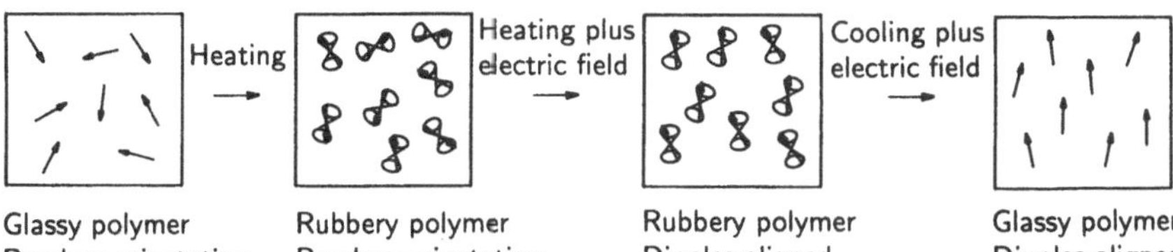

Fig. 8. Thermal-poling concept for glassy polymers with dipolar groups or guest molecules [89]

imum field and current can be controlled by adjusting the grid and corona voltages, and catastrophic electrical breakdowns can be avoided even in the presence of sample defects because of the local current limitation inherent in the corona process [98]. Corona poling at elevated temperatures is thus a suitable method for the preparation of optically active polymer films [99, 100]. In-situ optical measurements during thermal and room-temperature poling were employed for the investigation of the poling process itself [101, 102] and for the determination of optimal corona polarities and environments [103]; the highest possible stability was found with poling in helium [103]. Thermal poling of nematic polymers resulted in large nonlinear optical coefficients, not only parallel, but also perpendicular to the poling field [104].

Employing a ferroelectric copolymer of vinylidene fluoride and trifluoroethylene as host material led to improved stability of the dye-molecule orientation in guest-host systems poled at room temperature [94, 105]. Recently, chemical cross-linking of dye molecules and matrix under the electric field of the poling corona discharge at elevated or increasing temperatures led to nonlinear optical polymers that remain stable even at temperatures around 80°C [106, 107]. Liquid-crystalline side-chain polymers were recently found to exhibit an optically induced trans-cis isomerisation [108, 109] that may be used in optically addressed spatial light modulators and in high-density optical memories. In contrast to photoconductor-addressed systems, the electric field of the input light directly controls the changes in the electro-optic material and thus also the modulation of the output light. Spatial resolutions of 1500 line pairs per mm and diffraction efficiencies of up to 50% were already achieved with suitable materials [109]. In addition to second-order optical nonlinearities, third-order nonlinearities are also of interest; recently, relatively large such effects were found in a conjugated block copolymer [110]. Mechanisms and measurement techniques for third-order nonlinearities were investigated by means of quadratic electro-optic modulation [111].

Several methods are available for the determination of the electro-optic and nonlinear optical coefficients in polymer films: Beside more classical approaches deduced from the study of inorganic materials, novel experiments such as the use of an approximately 1-µm-thick Fabry-Perot interferometer containing the sample [112, 113], the measurement of waveguide properties with a prism-coupling technique [114], the determination of the phase retardation in the sample by means of a Babinet-Soleil compensator [115], and the detection of the rate of change of the relative phase shift between *p*- and *s*-polarised light [116] were introduced specifically for polymer materials.

Among the electro-optical applications suggested for polymer layers [91, 93, 94] is a transverse linear spatial light modulator [94, 117] as shown in Fig. 9. With such a demonstration device, a modulation depth of approximately 17% was achieved at a frequency of 8 kHz [117]. Longitudinal polymeric

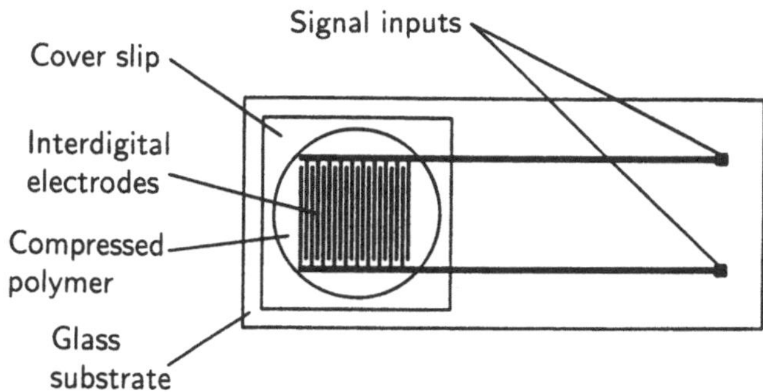

Fig. 9. Schematic drawing of a polymeric light modulator for the transverse linear electro-optic effect [117]

spatial light modulators were proposed for global optical interconnects between chips [118]; since optical beams can cross without interacting, high interconnection densities are possible, while the polymeric modulators may be applied directly to the surfaces of the respective microelectronic circuits [118].

5. Conclusions

Deformable light modulators with polymeric control layers offer the advantage of high diffraction efficiencies and, thus, also of high overall light fluxes, which are essential for projection devices. However, the optimal preparation and the long-term stability of such light valves is still under investigation. More recently introduced polymeric light modulators with two-component material systems such as polymer-dispersed liquid crystals and guest-host polymers look very promising. In view of their still rather limited contrast and efficiency capabilities, significant material improvements are necessary, and research in this direction is very active. Interest is fueled by the main advantages of polymer layers over competing materials, namely the ease of processing, the mechanical flexibility and durability, the possibility of molecular engineering for improved properties, and — in most cases — the full use of the available area for light modulation.

Acknowledgements

The author is indebted to W. Brinker, H.-J. Ehrke, N. Keil, Professor Dr. G. Mahler, W.-D. Molzow, Dr. G. Przyrembel, Dr. W. von Reden, H. Röder, T. Rosin, E. Schulze, Professor Dr. G. M. Sessler, T. Sinnig, Professor Dr. R. Tepe, W. Wirges, Professor Yao Hui-Hai, and S. Yilmaz for stimulating discussions and to the Bundesminister für Forschung und Technologie (BMFT) for financial support within the framework of the research projects Nos. TK 0444 4 and TK 0448 8; the responsibility for the contents of this paper rests, however, solely with the author.

References

1. Fisher AD (1990) Int J Optoelectron 5:125
2. Neff JA, Athale RA, Lee SH (1990) Proc IEEE 78:826
3. Baumann E (1952) J Brit Inst Radio Eng 12:69; (1953) J SMPTE 60:344
4. Glenn WE (1970) J SMPTE 79:788
5. Preston K (1969) Opt Acta 16:579
6. Preston K (1970) IEEE Trans Aerospace Electron Syst AES-6:458
7. Hornbeck LJ (1983) IEEE Trans Electron Devices ED-30:539
8. Pape DR, Hornbeck LJ (1983) Opt Eng 22:675
9. Pape DR (1985) Opt Eng 24:107
10. Hornbeck LJ (1990) Proc SPIE 1150:86
11. Fisher AD, Ling L-C, Lee JN, Fukuda RC (1986) Opt Eng 25:261
12. Rolsma PB, Lee JN, Oh T-K, Ling L-C (1989) Appl Opt 28:4816
13. Ling L-C, Fukuda RC, Fisher AD, Lee JN (1986) Proc SPIE 684:7
14. Rolsma PB, Lee JN (1990) Opt lett 15:712
15. Attard AE, Heffner BL (1981) Opt Lett 6:225
16. Attard AE (1982) Appl Opt 21:2348
17. Attard AE (1986) Appl Opt 25:2870
18. Attard AE, Kuehls JF (1984) Appl Phys Lett 44:522
19. Kuehls JF, Attard AE, Burke VB (1985) Appl Opt 24:3842
20. Glenn WE (1959) J Appl Phys 30:1870
21. Cressman PJ (1963) J Appl Phys 34:2327
22. Lee TC, Butter C (1979) Proc SPIE 202:147
23. Suemoto Y (1985) Proc SPIE 567:34
24. Moisan JY, Lever R, André B (1988) J Phys D: Appl Phys 21:513
25. Doyle RJ, Glenn WE (1971) IEEE Trans Electron Dev ED-18:739
26. Friesem AA, Katzir Y, Rav-Noy Z, Sharon B (1980) Opt Eng 19:659
27. Ineichen B, Liegeois C, Meyrueis P (1982) Appl Opt 21:2209
28. Lebreton G, Bamler R, Glünder H, Platzer H (1985) Appl Opt 24:450
29. Mast F, Baumgartner W, Held F, Baumann E (1955) Arrangement for Amplifying the Light Intensity of an Optically Projected Image. Swiss Pat 301222; (1959) US Pat 2896507
30. Baumgartner W (1967) Z Angew Math Phys 18:31
31. Sansom A, Kozol ET (1971) Deformographic Storage Display Tube (DSDT) — A Light-Valve Projection Display Having Controlled Persistence. In: Displays, IEE Conf Publ 80:325
32. Ross BJ, Kozol ET (1973) Performance Characteristics of the Deformographic Storage Display Tube (DSDT). In Proc 1973 IEEE Intern Conv Expos, IEEE, New York, 5:26/3
33. Sheridon NK (1972) IEEE Trans Electron Dev ED-19:1003
34. Sheridon NK (1973) The Ruticon as a Projection Display Device. In Proc 1973 IEEE Intern Conv Expos, IEEE, New York, 5:26/4
35. Lakatos AI (1974) J Appl Phys 45:4857
36. Bergen RF (1975) Appl Opt 14:1770
37. Lakatos AI (1975) J Appl Phys 46:1744
38. Sheridon NK, Berkovitz MA (1976) SPIE Proc 83:68
39. Kermisch D (1976) Appl Opt 15:1775
40. Lakatos AI (1977) J Appl Phys 48:2346
41. Lakatos AI, Bergen RF (1977) IEEE Trans Electron Dev ED-24:930

42. Wysocki JJ (1982) Appl Opt 21:2205
43. Bernstein H (1986) M Sc Thesis, Weizmann Institute of Science, Rehovot, Israel
44. Glenn WE (1987) SID Digest 87:72
45. Tepe R, Gerhard-Multhaupt R, Brinker W (1986) Proc SPIE 684:20
46. Tepe R (1987) J Opt Soc Am A 7:1273
47. Tepe R (1988) ntz-Archiv 10:269 and 10:295
48. Brinker W, Gerhard-Multhaupt R, Molzow W-D, Tepe R (1989) Proc SPIE 1018:79
49. Tepe R, Gerhard-Multhaupt R, Brinker W, Molzow W-D (1989) Appl Opt 28:4826
50. Gerhard-Multhaupt R, Brinker W, Tepe R (1989) Progr Colloid Polym Sci 80:63
51. Gerhard-Multhaupt R, Brinker W, Ehrke H-J, Molzow W-D, Roeder H, Rosin T, Tepe R (1990) Proc SPIE 1255:69
52. Azovtsev VP, Golosnoi OV, Shetakov AV, Gubanov IV, Kostyuk AV (1989) Instrum Exp Tech 32:679
53. Guscho YP (1991) Reliefography. Unpublished manuscript
54. Hess K, Dändliker R, Thalmann R (1987) Opt Eng 26:418
55. Craighead HG, Cheng J, Hackwood S (1982) Appl Phys Lett 40:22
56. Fergason JL (1985) SID Digest 85:68
57. Doane JW, Vaz NA, Wu B-G, Zumer S (1986) Appl Phys Lett 48:269
58. Fergason JL, Dalisa A, Lu S, Drzaic P (1986) SID Digest 86:126
59. Drzaic PS (1986) J Appl Phys 60:2142
60. Zumer S, Doane JW (1986) Phys Rev A 34:3373
61. Montgomery Jr GP, Vaz NA (1987) Appl Opt 26:738
62. Vaz NA, Montgomery Jr GP (1987) J Appl Phys 62:3161
63. Wu B-G, West JL, Doane JW (1987) J Appl Phys 62:3925
64. Golemme A, Zumer S, Doane JW, Neubert ME (1988) Phys Rev A 37:559
65. Montgomery Jr GP (1988) J Opt Soc Am B 5:774
66. Zumer S (1988) Phys Rev A 37:4006
67. Pirs J, Zumer S, Blinc R, Doane JW, West JL (1988) SID Digest 88:227
68. Vilfan M, Rutar V, Zumer S, Lahanjar G, Blinc R, Doane JW, Golemme A (1988) J Chem Phys 89:597
69. Drzaic PS (1988) Liquid Cryst 3:1543
70. Zumer S, Golemme A, Doane JW (1989) J Opt Soc Am A 6:403
71. Vaz NA, Montgomery Jr GP (1989) J Appl Phys 65:5043
72. Montgomery Jr GP, Vaz NA (1989) Phys Rev A 40:6580
73. Takizawa K, Kikuchi H, Fujikake H, Okada M (1990) Appl Phys Lett 56:999
74. Sansone MJ, Khanarian G, Leslie RM, Stiller M, Altmann J, Elizondo P (1990) J Appl Phys 67:4253
75. Welsh L, White L (1990) SID Digest 90:220
76. Kunigita M, Hirai Y, Ooi Y, Niiyama S, Asakawa T, Masumo K, Kumai H, Yuki M, Gunjima T (1990) SID Digest 90:227
77. Lauer H-U, Lueder E, Dobler M, Schleupen K, Spachmann J, Kallfass T, Jones P, Macknick B (1990) SID Digest 90:534
78. Pirs J, Olenik M, Marin B, Zumer S, Doane JW (1990) J Appl Phys 68:3826
79. Simoni F, Cipparrone G, Umeton C (1990) Appl Phys Lett 57:1949
80. Hirai Y, Niiyama S, Kumai H, Gunjima T (1990) Proc SPIE 1257:2
81. Vaz NA, Smith GW, Montgomery Jr GP (1990) Proc SPIE 1257:9
82. Cristiansen C (1884) Ann Phys 23:298; (1885) Ann Phys 24:439
83. Khanarian G, Tonelli AE (1983) Nonlinear Electro-optic and Dielectric Properties of Flexible Polymers. In: Williams DJ (ed) Nonlinear Optical Properties of Organic and Polymeric Materials. ACS Symp Ser 233:235
84. Williams DJ (1984) Angew Chem Int Ed Engl 23:690
85. Kowel ST, Ye L, Zhang Y (1985) Proc SPIE 567:44
86. Thakur M, Tripathy S (1986) Electrooptical Applications. In: Mark HF, Bikales NM, Overberger CG, Menges G, Kroschwitz JI (eds) Encyclopedia of Polymer Science and Engineering, Volume 5. John Wiley & Sons, New York, pp 756—771
87. Kowel ST, Ye L, Zhang Y, Hayden LM (1987) Opt Eng 26:107
88. Williams DJ (1987) Nonlinear Optical Properties of Guest-Host Polymer Structures. In: Chemla DS, Zyss J (eds) Nonlinear Optical Properties of Organic Molecules and Crystals, Volume 1. Academic Press, Orlando, San Diego, New York, London, pp 405—435
89. Singer KD, Lalama SL, Sohn JE, Small RD (1987) Electro-Optic Organic Materials. In: Chemla DS, Zyss J (eds) Nonlinear Optical Properties of Organic Molecules and Crystals, Volume 1. Academic Press, Orlando, San Diego, New York, London, pp 437—468
90. Khanarian G, Che T, DeMartino RN, Haas D, Leslie T, Man HT, Sansone M, Stamatoff JB, Teng CC, Yoon HN (1987) Proc SPIE 824:72
91. Stegemann GI, Seaton CT, Zanoni R (1987) Thin Solid Films 152:231
92. Prasad PN (1987) Thin Solid Films 152:275
93. Kowel ST, Selfridge R, Eldering C, Matloff N, Stroeve P, Higgins BG, Srinivasan MP, Coleman LB (1987) Thin Solid Films 152:377
94. Pantelis P, Hill JR, Oliver SN, Davies GJ (1988) Br Telecom Technol J 6:5
95. Lee C, Haas D, Man H-T, Mechensky V (1989) Photon Spectra 23:169
96. Möhlmann GR (1990) Europhys News 21:83
97. Singer KD, Sohn JE, Lalama SJ (1986) Appl Phys Lett 49:248
98. Sessler GM (ed) (1987) Electrets, 2nd Enlarged Edition. Springer-Verlag, Berlin, Heidelberg, New York
99. Singer KD, Kuzyk MG, Holland WR, Sohn JE, Lalama SJ, Comizzoli RB, Katz HE, Schilling ML (1988) Appl Phys Lett 53:1800

100. Mortazavi MA, Knoesen A, Kowel ST, Higgins BG, Dienes A (1989) J Opt Soc Am B 6:733
101. Page RH, Jurich MC, Reck B, Sen A, Twieg RJ, Swalen JD, Bjorklund GC, Willson CG (1990) J Opt Soc Am B 7:1239
102. Eich M, Sen A, Looser H, Bjorklund GC, Swalen JD, Twieg R, Yoon DY (1989) J Appl Phys 66:2559
103. Hampsch HL, Torkelson JM, Bethke SJ, Grubb SG (1990) J Appl Phys 67:1037
104. Yitzchaik S, Berkovic G, Krongauz V (1990) Opt Lett 15:1120
105. Hill JR, Dunn PL, Davies GJ, Oliver SN, Pantelis P, Rush JD (1987) Electron Lett 23:701
106. Eich M, Reck B, Yoon DY, Willson CG, Bjorklund GC (1989) J Appl Phys 66:3241
107. Jungbauer D, Reck B, Twieg R, Yoon DY, Willson CG, Swalen JD (1990) Appl Phys Lett 56:2610
108. Wendorff JH, Eich M (1989) Mol Cryst Liq Cryst 169:133
109. Eich M, Wendorff JH (1990) J Opt Soc Am B 7:1428
110. Jenekhe SA, Chen W-C, Lo S, Flom SR (1990) Appl Phys Lett 57:126
111. Kuzyk MG, Sohn JE, Dirk CW (1990) J Opt Soc Am B 7:842
112. Uchiki H, Kobayashi T (1988) J Appl Phys 64:2625
113. Eldering CA, Kowel ST, Knoesen A (1989) Appl Opt 28:4442
114. Horsthuis WHG, Krijnen GJM (1989) Appl Phys Lett 55:616
115. Teng CC, Man HT (1990) Appl Phys Lett 56:1734
116. Schildkraut JS (1990) Appl Opt 29:2839
117. Hill JR, Pantelis P, Abbasi F, Hodge P (1988) J Appl Phys 64:2749
118. Eldering CA, Kowel ST, Mortazavi MA, Brinkley PF (1990) Appl Opt 29:1142

Author's address:

Dr. R. Gerhard-Multhaupt
Heinrich-Hertz-Institut für Nachrichtentechnik
Einsteinufer 37
1000 Berlin 10, FRG

Third harmonic generation in perylene derivatives

S. Schrader*), K. H. Koch, A. Mathy, C. Bubeck, K. Müllen, and G. Wegner

Max-Planck-Institut für Polymerforschung, Mainz, FRG
*) Zentralinstitut für Organische Chemie, Bereich Makromolekulare Verbindungen, Berlin, FRG

Abstract: The nonlinear optical properties of a series of tetra-tert.-butyl-perylene derivatives of different molecular sizes were investigated by means of third harmonic generation. — The nonlinear susceptibilities $\chi^{(3)}(-3\omega, \omega, \omega, \omega)$ of the smaller molecules (biperylenyle, terrylene, quaterrylene) as measured by excitation with a Nd:YAG-laser (wavelength = 1064 nm) are in the range of 10^{-12} esu and are one order of magnitude lower than that of the pentarylene. Its $\chi^{(3)}$-value reaches a magnitude greater than 5×10^{-11} esu which is comparable to the $\chi^{(3)}$ of linear conjugated polymers. The nonlinear susceptibility of the investigated perylene derivatives increases with the interrelated quantities of the reciprocal of the $S_0 \rightarrow S_1$-excitation energy, the wavelength of the absorption maximum, and the length of the molecule. Resonance enhancement of the nonlinear susceptibility is observed for the biperylenyle due to a mutual two-photon resonance and for the pentarylene due to both one-photon and three-photon resonance.

Key words: Third harmonic generation; nonlinear optical properties; perylene derivatives

Introduction

Conjugated organic polymers show high values of the nonlinear optical susceptibility $\chi^{(3)}(-3\omega, \omega, \omega, \omega)$ due to their highly polarizable π-electron system [1]. Especially linear conjugated polymers show a power-law dependence of third-order susceptibility on the wavelength of the absorption maximum [2], similar to the scaling of the nonlinear susceptibility with the sixth power of the reciprocal gap energy E_g^{-6}, as predicted by Flytzanis et al. for one-dimensional semiconductors [3].

The connection between nonlinear optical susceptibility and the size and dimension of the π-electron system is of fundamental importance for a detailed understanding of the nonlinear optical properties of organic molecules.

The present study of third harmonic generation (THG) on a series of perylene derivatives of different molecular size contributes especially to this topic. The investigated compounds are tetra-tert.-butyl-oligo(rylene)s with the chemical structures given in Fig. 1. First results of THG investigations of these compounds have been described recently [4]. The first member of this series, perylene, has been thoroughly investigated. The synthesis of the higher homologues, the unsubstituted terrylene and quaterrylene were described in the 1950s [5]; however, due to their extreme insolubility in organic solvents, an adequate characterization has not yet been possible. Due to the substitution with tetra-tert.-butyl-groups, the higher rylenes are still soluble and, hence, a complete characterization was possible [6—8]. Thus, they are valuable model compounds for a systematic study of nonlinear optical properties of planar polyconjugated materials.

Experimental

The investigated materials are the following perylene derivatives: tetra-tert.-butyl-biperylenyle (TTBBP), tetra-tert.-butyl-terrylene (TTBT), tetra-tert.-butyl-quaterrylene (TTBQ), and tetra-tert.-butyl-pentarylene (TTBP) [6—9]. Their chemical structures are given in Fig. 1. Synthesis and characterization of the substances are described in the papers of Koch et al. [6—9]. The investigated films were prepared by vacuum sublimation on fused silica.

TTBPer
TTBBP
TTBT
TTBQ
TTBP

Fig. 1. Chemical structure of different tetra-tert.-butyl-(TTB)-oligo-(rylene)s [6—9]: TTBPer = perylene; TTBBP = biperylenyle; TTBT = terrylene; TTBQ = quaterrylene; TTBP = pentarylene

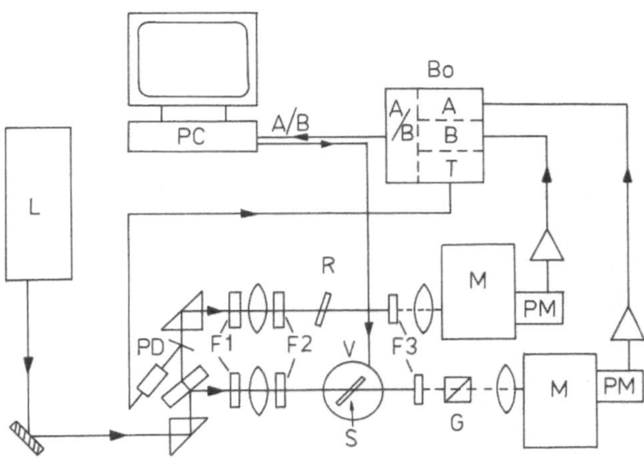

Fig. 2. Experimental set-up for measurements of the third harmonic generation (THG) in thin organic layers: L = Nd:YAG-laser, 1064 nm, 0.4 mJ/pulse, 35 ps, 10 Hz; PD = photodiode; F1 = neutral filters; F2 = filters for blocking harmonic light; F3 = filters for blocking the fundamental light; R = reference; V = vacuum chamber; S = sample rotated by a step-motor; G = Glan-Taylor-polarizer; M = monochromator; PM = photomultiplier; A = amplified signal from the sample; B = amplified reference signal; Bo = two-channel boxcar-integrator and analog device for division A/B; T = trigger; PC = personal computer

We obtained soft homogeneous films of 200—300-nm thickness. In the case of substituted perylene (TTBPer), the films had large scattering losses due to its strong tendency towards crystallization. Therefore, we investigated the TTBBP molecule, which is electronically very close to the substituted perylene, but which forms homogeneous, transparent evaporation layers.

The films were characterized by UV/VIS/NIR-, IR-, and Ramanspectroscopy. Their thicknesses were determined by a step-profiler.

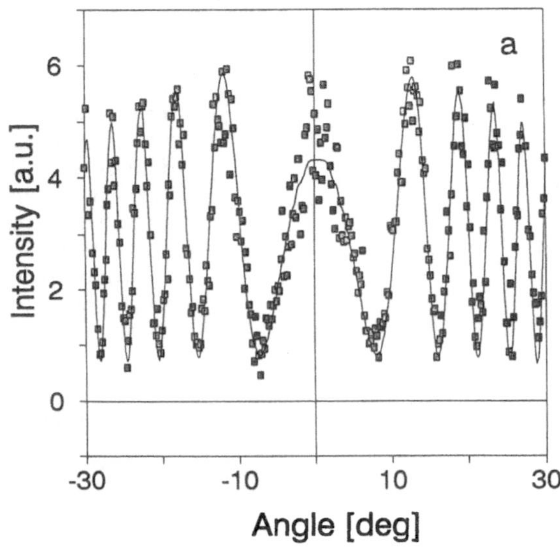

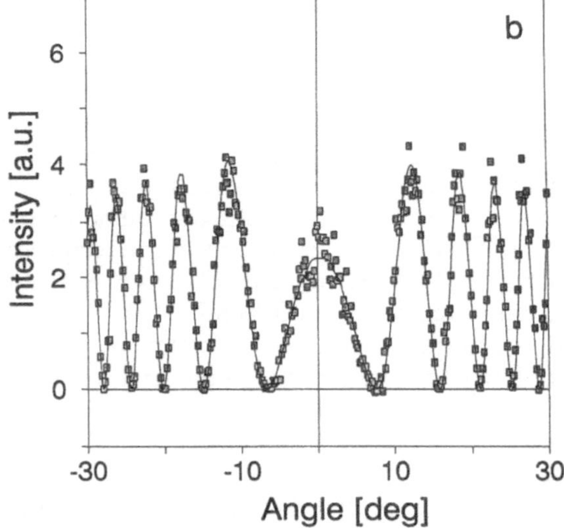

Fig. 3. Maker-fringes as measured in the THG-experiment a) fused silica plate of 1-mm thickness evaporated with 290-nm thick tetra-tert.-butyl-biperylenyle layer; b) the same plate, at the same position, but cleaned of the evaporated layer

The experimental set-up for the THG-measurements was described recently [10—12]; it is shown schematically in Fig. 2. The incident laser beam of a Nd:Yag-laser is split into reference and measuring beams. In the reference beam the THG-signal of a reference material, and in the measuring beam the THG-signal of the sample are generated and detected by using a monochromator and a photomultiplier. After integration in a boxcar-integrator the sample signal is divided by the reference signal in order to compensate the influence of intensity fluctuations of the laser pulses. Subsequently, the corrected value of the sample signal is recorded by a personal computer which runs the whole experiment automatically [10]. During the experiment the sample is placed in a vacuum chamber in order to avoid THG-signals generated by air.

The detection method of THG is based on the Maker-fringe technique [13]. The sample is rotated in the laser beam, and in dependence on the incident angle maxima and minima of the harmonic light are detected. This can be seen in Figs. 3a and 3b, where the Maker-fringe pattern of a fused silica plate with evaporated TTBBP, as well as the pattern of the same plate at the same position after removing the evaporated layer are presented. The cleaned fused silica plate (Fig. 3b) serves as a reference which is investigated under identical experimental conditions as the material of interest.

The detected third harmonic intensity was analyzed via a formalism including all bound waves as described in [10—12]. The absolute value and phase angle of the complex quantity $\chi^{(3)}(-3\omega,\omega,\omega,\omega)$ were evaluated with respect to the reference value of fused silica ($\chi^{(3)} = 3.11 \cdot 10^{-14}$ esu).

Results and discussion

The absorption spectra of the evaporated rylene films are given in Fig. 4. These spectra represent the lowest electronic transition ($S_0 \rightarrow S_1$) which appears between 450 nm and 800 nm, depending on the size of the molecule. The absorption coefficients α_{max} at the wavelength of the absorption maximum λ_{max} are in the order of 10^5 cm^{-1}. As can be seen in Fig. 4, the spectra show vibronic sidebands, which are separated in energy by a value of about 0.17 eV. This corresponds to aromatic C—C-vibrations of the molecule, as can be concluded from Raman-, and IR-measurements. Higher energetic transitions of the order equal to or greater than the $S_0 \rightarrow S_4$-transition are responsible for absorption at wavelengths below 400 nm, as was confirmed by model calculations using the Pariser-Parr-Poples (PPP) method [14].

Table 1 presents the linear and nonlinear optical properties as received from our measurements on the solid rylene layers. There, the absolute value

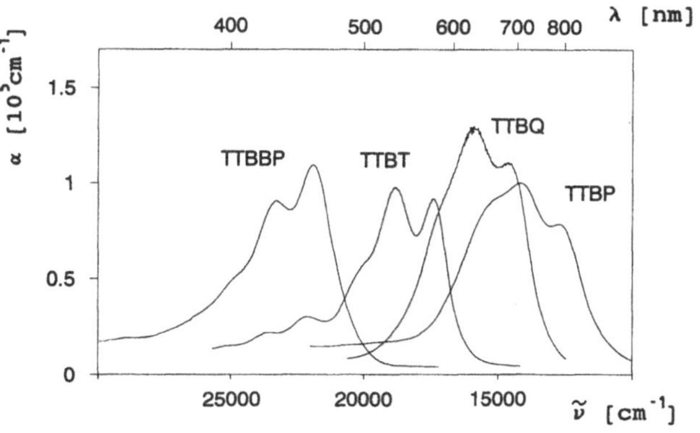

Fig. 4. Absorption spectra of different oligo(rylene)s with the structures shown in Fig. 1 [4]

$\chi^{(3)}$ and the phase angle Φ of the complex nonlinear third-order susceptibility $\chi^{(3)}(-3\omega,\omega,\omega,\omega)$ are shown. In addition, the linear absorption coefficient α_{max}, as well as the linear refractive indices $n(\omega)$ and $n(3\omega)$ at the fundamental and at the harmonic frequency, respectively, are given. The refractive indices $n(\omega)$ and $n(3\omega)$ were estimated by Kramers-Kronig analysis of the absorption spectra and, based on these values, the evaluation of $\chi^{(3)}$ and Φ was carried out.

The absolute values of $\chi^{(3)}$ of the shorter rylenes (TTB-biperylenyle, TTB-terrylene, TTB-quaterrylene) of about 10^{-12} esu are more than one order of magnitude lower than that of TTB-pentarylene, and lower than the values of many linear conjugated polymers [1, 2].

The phase angles Φ for TTBBP and for TTBP indicate resonance enhancement of the nonlinear susceptibility.

Because there is no linear absorption at 532 nm for the TTBBP the high contribution of the imaginary part to the nonlinear susceptibility should be due to a two-photon absorption process.

In the case of TTBP the low-energy-tail of the absorption connected with the $S_0 \rightarrow S_1$-transition is already influenced by the excitation wavelength of 1064 nm. Therefore, the one-photon-resonance contributes to the imaginary part of the nonlinear susceptibility. On the other hand, a weak absorption band below 400 nm which is only present for the pentarylene is the reason for a three-photon resonance which should strongly contribute to the imaginary part and, hence, to the high, resonance enhanced $\chi^{(3)}$-value of TTBP. In fact, from PPP-calculations for TTBP it follows that the $S_0 \rightarrow$

Table 1. Nonlinear and linear optical properties of different tetra-tert.-butyl-oligo(rylene)s as described in the text; ω respresents the frequency of the fundamental wavelength, λ = 1064 nm of a Nd:YAG-laser

Rylenes	$\chi^{(3)}$ [10^{-12} esu]	Φ [deg]	λ_{max} [nm]	$n(\omega)$	$n(3\omega)$	α_{max} [10^5 cm^{-1}]
TTBBP	0.48	263	457	1.72	1.59	1.19
TTBT	0.81	153	574	1.81	1.66	0.99
TTBQ	1.57	5	689	2.15	1.77	1.19
TTBP	80	310	785	2.9	1.6	0.85

S_4-transition has a considerable oscillator strength unlike the smaller rylenes and should, therefore, be responsible for the absorption peak between 340 and 410 nm. Due to the poor quality of the sublimated TTBP-films the phase angle was determined less accurately than for the other materials.

The phase angles of 153° and 5° for TTBT and TTBQ, respectively, underscore the nonresonant character of their nonlinear susceptibility under the chosen excitation conditions.

The $\chi^{(3)}(-3\omega, \omega, \omega, \omega)$-values of the investigated rylenes increase systematically with the maximum wavelength of absorption. This is in qualitative agreement with earlier observations on other dye systems [15] and with the above-mentioned predictions of Flytzanis et al., where the nonresonant, nonlinear third-order susceptibility of one-dimensional semiconductors scales with the sixth power of the inverse gap energy [3]. In the case of the observed amorphous organic layers, the nonresonant $\chi^{(3)}$ values should scale with a power-law dependence of the inverse of the $S_0 \rightarrow S_1$-excitation energy and, consequently, with the position of the absorption maximum.

A direct comparison with the properties of the well-studied linear, one-dimensional polyconjugated chains such as polyenes and polydiacetylenes is not possible, because the oligo-rylenes are not really one-dimensional systems. They exhibit a different dimensionality towards two-dimensional behavior.

Acknowledgement

The authors thank Mr. H. Menges for substantial technical support. Financial support for this work was given by the BMFT under project number 03M4008E9.

References

1. Chemla DS, Zyss J (eds) (1987) Nonlinear Optical Properties of Organic Molecules and Crystals. Academic Press, New York
2. Neher D, Kaltbeitzel A, Wolf A, Bubeck C, Wegner G (1990) In: Bredas JL, Chance RR (eds) Conjugated Polymeric Materials — Opportunities in Electronics. Kluwer Acad Publ, Dordrecht, p 387
3. Agrawal GP, Cojan C, Flytzanis C (1987) Phys Rev B17:776
4. Schrader S, Koch KH, Mathy A, Bubeck C, Müllen K, Wegner G (1991) Synth Met 43/1—2:3223, Fig. 4 by permission of Elsevier Sequoia S. A. Publishers, Lausanne
5. Clar E, Kelly W, Laird RM (1956) Mh Chem 87:391
6. Bohnen A, Koch KH, Lüttke W, Müllen K (1990) Angew Chem Int Ed Engl 29:525
7. Koch KH, Fahnenstich U, Baumgarten M, Müllen K (1991) Synth Met 41—43
8. Koch KH (1991) Thesis. Johannes-Gutenberg-University Mainz, Germany
9. Koch KH, Müllen K (1991) Chem Ber: submitted
10. Neher D (1990) Thesis. Johannes-Gutenberg University Mainz, Germany
11. Neher D, Wolf A, Kaltbeitzel A, Bubeck C, Wegner G (1991) J Phys D Appl Phys: in press
12. Neher D, Wolf A, Bubeck C, Wegner G (1989) Chem Phys Lett 163:116
13. Maker PD, Terhune RW, Nisenoff M, Savage CM (1962) Phys Rev Lett 8:21
14. Koch KH (1990): unpublished results
15. Hermann JP, Decuing J (1974) J Appl Phys 45:5100

Authors' address:

Dr. S. Schrader
Zentralinstitut für Organische Chemie
Bereich Makromolekulare Verbindungen
Rudower Chaussee 5
O-1199 Berlin, FRG

Discussion

SWORAKOWSKI:

One of the basic requirements for the nonlinear materials to be applied for the generation of higher harmonics is that they do not adsorb at the wavelengths of the incident radiation, nor at harmonic frequencies. In the case of your materials this requirement seems not to be fulfilled. Can you comment on that?

SCHRADER:

The aim of our investigations is to contribute to the basic understanding of the relationship between molecular structure and their nonlinear optical third-order effects. In this case, the pecularities of the π-electron system of such planar molecules were studied by varying the molecular size in a systematic manner. In the case of our excitation conditions with a fundamental wavelength of 1064 nm, we had, indeed, considerable resonance enhancement for the biggest molecule, the pentarylene. From the knowledge about the change of the absorption properties with molecular size, one should design the active molecules for practical purposes in a way that the one-photon absorption is energetically far enough above the wavelength of the excitation source. On the other hand, two-, and especially three-photon absorption processes of the molecule should also be outside the second and third harmonic of the source. Otherwise, the material would be chemically degraded under the impact of the high-intensity laser beam. A compromise in designing molecules for lower excitation light intensities consists in modifying the molecule so that the tails of these absorption processes are shifted closer to the excitation wavelength or its higher harmonics in order to use a certain resonance enhancement to gain sufficient intensities of the generated harmonic light.

WENDORFF:

What approach can be taken to achieve phase matching for the systems you described in your lecture?

SCHRADER:

The easiest way is to choose a wave-guide structure of proper geometry for such nonlinear optical materials and, if the film-forming properties are poor, to incorporate them in appropriate, film-forming polymers.

Dielectric and thermal relaxations in low molecular mass liquid crystals

C. Schick*), B. Stoll**), J. Schawe*), A. Roger*), and M. Gnoth**)

*) Hochschule Güstrow, Institut für Physik, Güstrow, FRG
**) Universität Ulm, Abteilung Angewandte Physik, FRG

Abstract: The investigation of amorphous layers between crystalline lamellae allows the estimation of glass transition length scales. Low molecular mass liquid crystals can form semicrystalline structures without extensive interfacial layers between crystalline and liquid regions like in polymers. First results from dielectric spectroscopy and calorimetry are presented. The semicrystalline sample shows a, complicated relaxation behavior without a typical glass transition. A thermally activated local process is observed (f_0 = 10^{20} Hz; E_A = 118 kJ/mol). It may be a residuum of the glass transition. Further investigations are necessary to get detailed information about structure and relaxation behavior of these interesting materials.

Key Words: Liquid crystal; thermal behavior; dielectric relaxation; activation parameters; glass transition; spatial limitations

Introduction

Glass transition is a universal phenomenon [1]. It can be observed, not only in ordinary silicate glasses, but also in low molecular weight organic liquids [2], salt melts [3], polymers (e.g., [4]), liquid and plastic crystals (e.g., [5]), and other substances with disorder.

Predictions about typical length scales associated with the glass transition from different models (e.g., [6—9]) results in values of some nanometers. A direct experimental verification in the equilibrium is, at present, not possible. Therefore, we must try to use indirect experimental methods to estimate the characteristic length scales. As the glass transition is a universal phenomenon with respect to these predictions, we can use the advantage of a relative free choice of testing substances.

The experiments on semicrystalline poly(ethylene terephthalate) [10—12] show significant sensitivity of the glass transition parameters to the nanometer length scale. The results can be explained by a direct influence of spatial limitations on the large modes of the glass transition. We found that there is an actual correlation between the layer thickness of the amorphous region and the characteristic length of glass transition that we get from Donth's fluctuation model [6]. Figure 1 shows the central result of our work on PET [10].

Further discussion should implement the following appearances concerned with the structure formation of semicrystalline polymers:

i) There are not only crystalline and melt-like amorphous layers, but also an interfacial region between them; however, not enough is known about the relaxation behavior of these interfacial regions.
ii) It is not possible to form thinner or thicker amorphous layers inside a stack of PET-lamellae, as shown in Fig. 1.

Therefore, we carried out investigations in semicrystalline low molecular mass systems because there is not such an interfacial region due to the folding of the polymer chains, and it should be possible to form thinner or thicker amorphous regions between the crystallites (no lamellaes).

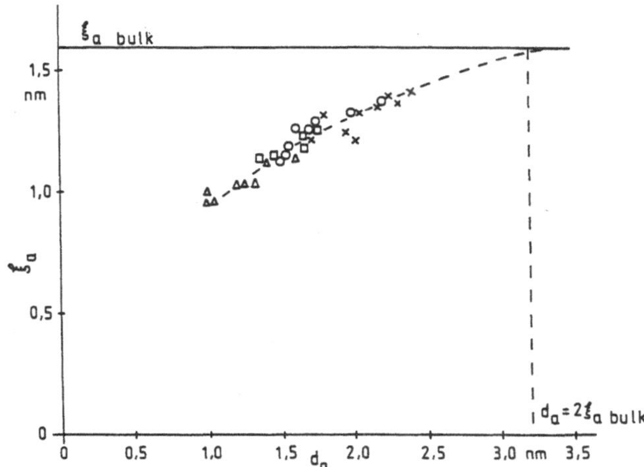

Fig. 1. Correlation length ξ_a (from thermal glass transition interval Eqs. (5), (6) [6]) as a function of amorphous layer thickness d_a (from SAXS, WAXS, and ΔC_p [11]) for semicrystalline PET prepared by different crystallization regimes. The dotted curve is the extrapolation expected towards bulk amorphous sample [10]

Experimental

The liquid crystal studied was 3,3′-sulphonyl-bis-[methyl 4-(4-n-pentyloxy-benzoyloxy)-benzoate] (Fig. 2) [13]. This compound shows a nematic phase with appears only when supercooling the isotropic melt, because the melting point (T_m = 443 K) is higher than the clearing point (T_{NI} = 432 K).

The appearance of liquid-crystalline properties is an interesting phenomenon for this molecular structure, because there are two mesogenic phenyl benzoates linked as rod-like rigid molecular halves by a sulphonyl group, forming a "siamese twin" mesogen.

The synthesis of this compound can be carried out by using a simple three-step procedure starting from methyl 4-hydroxy-benzoate [13]. The crude product was recrystallized from methanol five times. We obtained crystalline flat plates (about 1 · 0.5 · 0.1 mm^3). By means of several investigation methods the antiparallel conformation of both molecular halves in the netmatic state could be proved [13, 14].

The substance is an interesting model due to the following facts.

i) With supercooling of the melt, it forms a nematic glass with a glass temperature of 302 K. It is possible to do this with extremely low cooling rates down to 2.5 K/min. Thus, it was possible to perform experiments while cooling the sample from the melt (Fig. 3; Fig. 9).
ii) When crystallizing from an ethanol-solution, it forms a 100% crystalline structure. The melting point is at 442 K and the heat of fusion is about 60 J/g.
iii) Annealing at temperatures above T_g, the nematic liquid forms semicrystalline structures with crystallinities up to 95%, but not 100%. That means that there are small amorphous regions inside the sample, and we were able to investigate the relaxation behavior of these amorphous regions (Fig. 4).

In Fig. 5 it is demonstrated that these liquid crystalls crystallize actually represent a two-phase system. The line A stands for a two-phase system

Fig. 2. 3,3′-sulphonyl-bis[methyl 4-(4-n-pentyloxy-benzoyloxy)-benzoate]

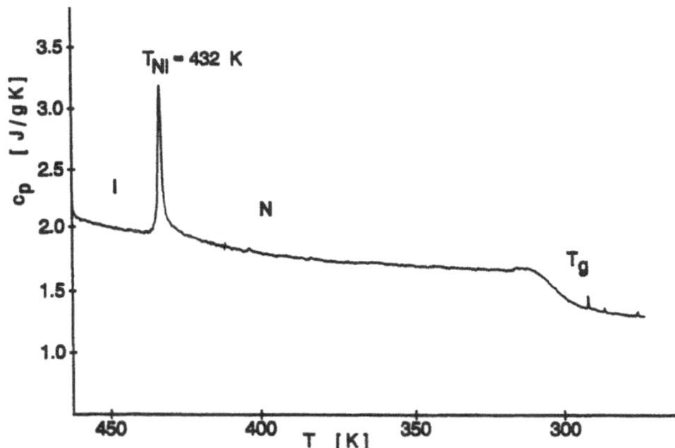

Fig. 3. Specific heat capacity of the liquid sample while cooling from the melt with a cooling rate of 10 K/min. *I* = isotropic liquid; T_{NI} = clearing point; *N* = nematic liquid; T_g = glass transition

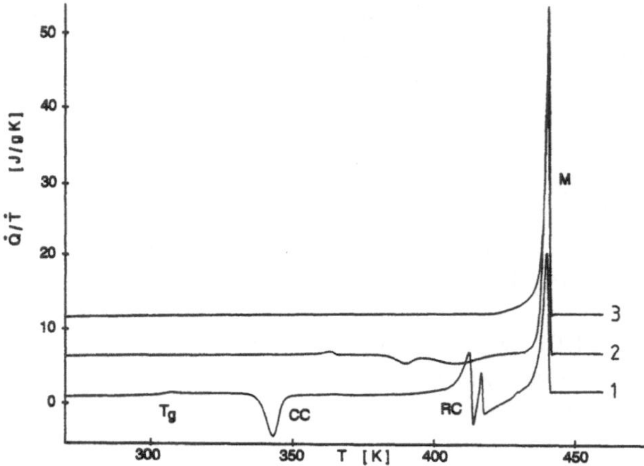

Fig. 4. Thermograms (heating rate 10 K/min) after cooling the sample at 10 K/min from the melt and annealing for different times at 310 K. A = 0 h; B = 3 h; C = 13 h. For the sake of clarity the successive curves are vertically shifted. T_g = glass transition; CC = cold crystallization; RC = recrystallization; M = melting

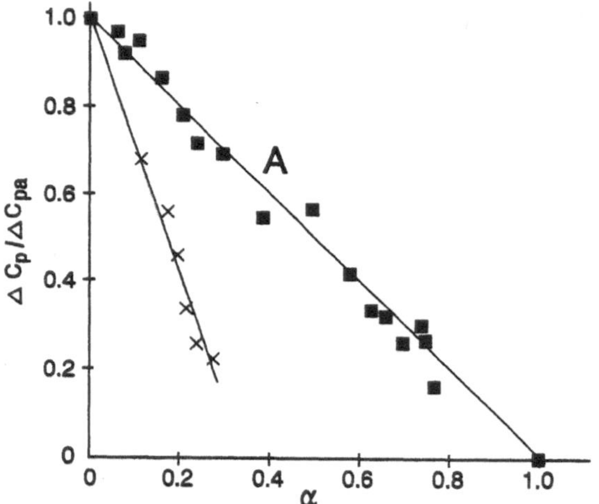

Fig. 5. Normalized step height of the specific heat capacity of the glass transition (ΔC_p = sample under investigation; ΔC_{pa} = liquid) as a function of crystallinity. ■ = liquid crystal; × = PET; isothermal crystallization at 390 K. The line A represents a two-phase behavior

where we have only the coexistence of the crystalline and the melt-like amorphous phases. The values from the liquid crystal coincide with this line. On the other hand, the values from the polymer (PET) show a significant deviation from the two-phase behavior due to the interfacial regions [15].

This liquid crystalline material allows us to investigate the relaxation behavior in the pure nematic and in the 100% crystalline state, as well as in a semicrystalline state containing only these two phases. In the following, we present first results from dielectric and thermal investigations.

For the measurements, we used the crystals from recrystallization from ethanol. A layer of about 0.3 mm was situated between the capacitor plates. Small teflon strips controlled the distance. The crystalline material was measured and then the sample was melted inside the capacitor and cooled down to room temperature to prepare a nematic glass. The sample was measured again and then annealed at 310 K for 13 h to produce a semicrystalline sample. For the caloric measurements the same procedure was performed inside the calorimeter.

The dielectric measurements were done with a modified Schering-bridge [16] and a Network-Analyser (HP 3577A) [17] at constant temperature (190 K < T < 373 K) in the frequency range from 10^{-2} Hz up to $5 \cdot 10^6$ Hz. Representative plots are shown in Figs. 6–8.

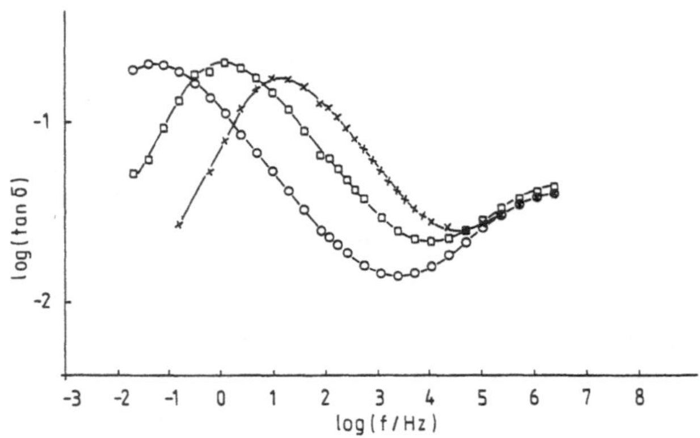

Fig. 6. Dielectric spectra of the nematic liquid at constant temperature. ○ = 303 K; □ = 307 K; × = 353 K

During cooling from the melt with 5 K/min, or heating, dielectric measurements were performed with an automatic bridge from Tetrahedron (USA) [18] at frequencies near 0.1; 1; 10 and 100 kHz. Figure 9 shows the curve of one amorphous sample, and Fig. 10 one of a semicrystalline sample.

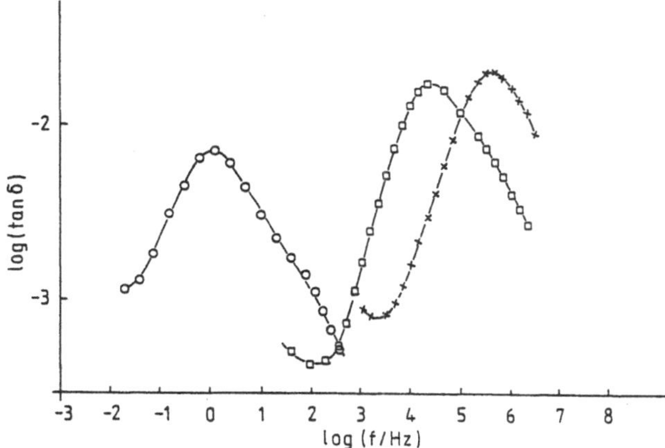

Fig. 7. Dielectric spectra of the 100% crystalline sample at constant temperature. ○ = 218 K; □ = 313 K; × = 353 K

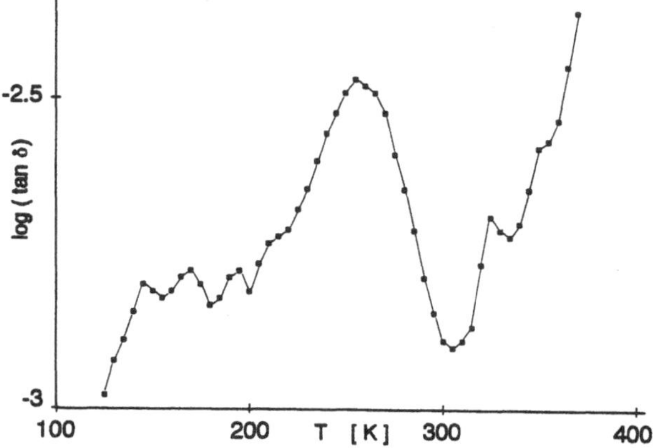

Fig. 10. Dielectric spectra of a semicrystalline sample (T_a = 310 K; t_a = 13 h) at constant frequency. f = 1 kHz; heating rate 5 K/min

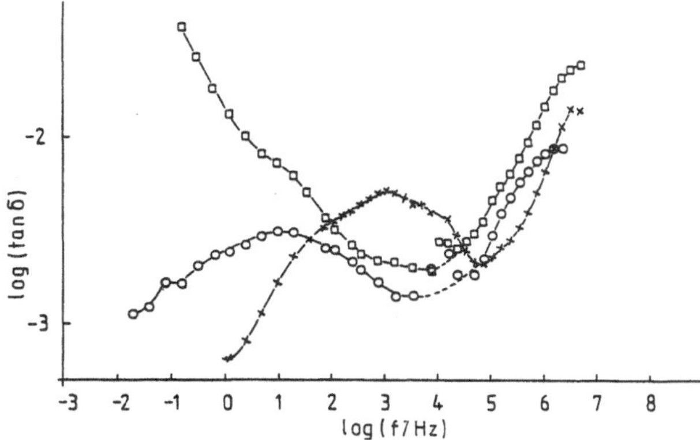

Fig. 8. Dielectric spectra of a semicrystalline sample (T_a = 310 K; t_a = 13 h) at constant temperature. ○ = 207 K; × = 242 K; □ = 340 K

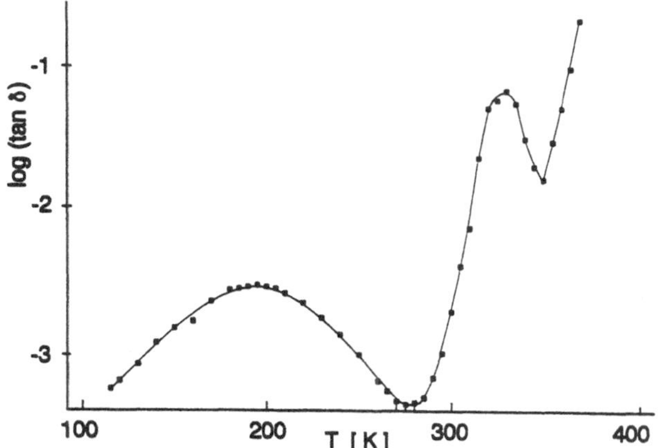

Fig. 9. Dielectric spectra of the nematic liquid while cooling from the melt at constant frequency. f = 1 kHz; cooling rate about 5 K/min

Because we do not know the correct geometry of the sample, especially of the crystalline powder, we were only able to determine the tanδ and its maximum.

The caloric measurements were performed with a DSC-2 (Perkin-Elmer). The heating rate was 10 K/min and the sample weight about 5 mg. From curve 1, Fig. 4, a glass-transition (T_g), cold crystallization (*CC*), recrystallzation (*RC*) and melting (*M*) can be recognized, such as we know from polymers like PET.

To compare the caloric with the dielectric measurements, the heating or cooling rates were transformed to the corresponding frequency by a formula from the dislocation concept of glass transition [19].

$$f \approx \dot{T}/(2 \cdot \pi \cdot 15 \text{ K}) \ . \tag{1}$$

This is a very rough approximation. The same order of magnitude gives a relation from the fluctuation model of glass transition [6]

$$f \approx \dot{T}/(2 \cdot \pi \cdot a \cdot \delta T) \ , \tag{2}$$

where $\dot{T}$ is the heating rate, δT the mean temperature fluctuation of about 2.6 K, and a a constant ($a \approx 19$ [20]).

To locate the thermal glass transition in the activation diagrams (Figs. 12, 15) the points (triangles) were determined by the frequencies calculated via (1), and the temperature of the half step of the specific heat capacity (glass temperature T_g) [19].

Results

The dielectric and the thermal measurements were used to construct the activation diagrams of the nematic liquid (Fig. 12), the 100% crystalline (Fig. 11) and one semicrystalline (Fig. 13) sample. Most of the points represent the maximum frequency of the $\tan\delta$ peak from dielectric measurements at constant temperature performed at the University of Ulm, FRG. With this technique, we can measure the liquid sample only up to temperatures of 320 K, because crystallization starts during the measurement at higher temperatures.

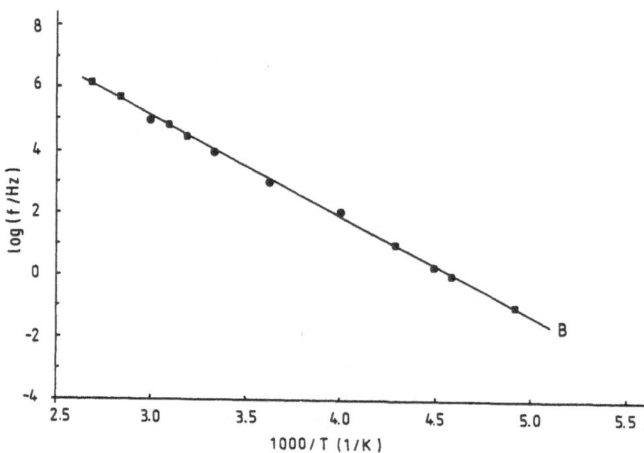

Fig. 11. Activation diagram of the 100% crystalline sample from dielectric spectra at constant temperature (■) and at constant frequency (●)

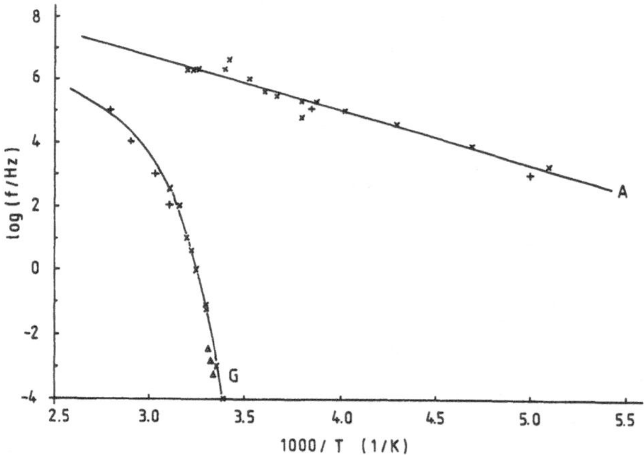

Fig. 12. Activation diagram of the nematic liquid from dielectric spectra at constant temperature (×); at constant frequency (+), and from the thermal glass transition (△)

Thus, it was necessary to measure at constant frequency and decreasing temperature (starting the experiment in the melt). Such measurements were done with the automatic bridge in Güstrow at frequencies near 0.1, 1, 10, and 100 kHz, from 460 K down to 110 K, and cooling rates of about 5 K/min. We were thus able to measure liquid samples down to the beginning of the thermal glass transition at about 310 K (Fig. 12 (+)).

In the activation diagram of the 100% crystalline sample (Fig. 11), we can only detect one local process, characterized by the Arrhenius-line B. That means that it is not possible to realize cooperative molecular motions in this material which are necessary for a glass transition.

Figure 12 shows a typical activation diagram of a glass-forming liquid. Table 1 shows the parameters of the local process (line A). Curve G elucidates the glass process; this curve was calculated according to:

$$\log(f_m) = \log(f_0/\pi) - Q/(2.303\,RT) + 3(3r/d)^2 d/s \log(1 - (1 - x)^{3r/d}) \quad (3)$$

$$x = \exp(-\varepsilon_s/RT)\,, \quad (4)$$

where x is the concentration, of segment dislocations according to the dislocation concept (for details see [7, 19]). The units of cooperative motion in this model are cubes with the edge length $3r$. This model was applied, not only to polymers, but also to liquid crystals [21]. Equation (3) was fitted to the experimental points in Fig. 12 using the following parameters:

Q (local activation energy) = 49 kJ/mol;
σ_s (free energy of segment dislocations) = 4.465 kJ/mol;
$3r/d$ (d = molecular distance) = 37.

The other parameters (not fitted) were:

f_0 (local vibration frequency) = 10^{12} Hz;
ζ (intermolecular part of Q) = 0.6,

and the coefficient for the temperature dependence of ε_s was assumed to be known.

As described in [6, 22], the size of cooperatively rearranging regions ξ_a can also be estimated from the mean temperature fluctuation of the cooperative subsystem (δT), according to

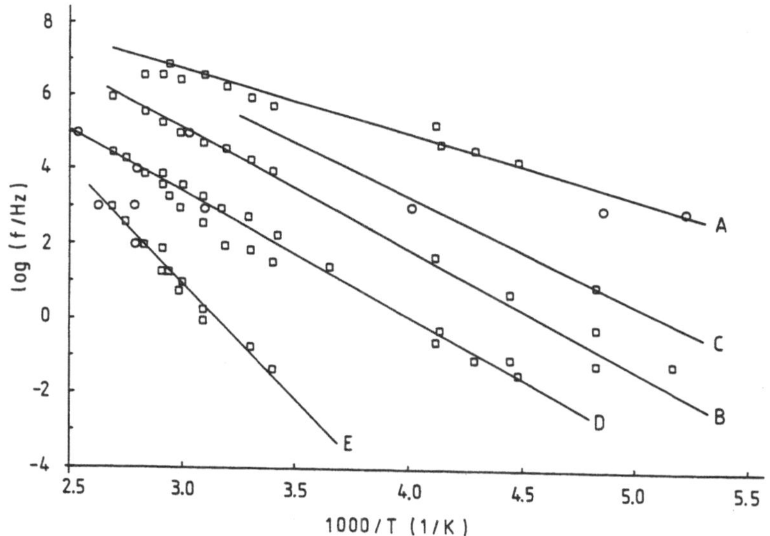

Fig. 13. Activation diagram of a semicrystalline sample (T_a = 310 K; t_a = 13 h) from dielectric spectra at constant temperature (□) and at constant frequency (○)

Table 1. Parameters describing the local processes in Fig. 15, following the equation $f_m = f_0 \exp(-E_A/R \cdot T)$

Process	$\lg(f_0/\text{Hz})$	$E_A/\text{kJ mol}^{-1}$
A	12	34
B	14	58
C	14	53
D	13	63
E	20	118

$$V_a = kT_g^2 \Delta(1/C)_v/(\delta T^2 \rho) \quad (5)$$

$$\xi_a = (3V_a/4\pi)^{1/3} , \quad (6)$$

where ρ is the bulk density, $\Delta(1/C_v)$ is the step of the reciprocal specific heat capacity near T_g, and ξ_a is the corresponding "radius" of the cooperatively regions characterized by the volume V_a. δT can be extracted from the glass transition interval ($\delta T \approx 0.4\Delta T$; $\Delta T \approx$ 6 K from Fig. 14; $\longrightarrow \delta T \approx$ 2.4 K). For details see [6, 22]. The result for ξ_a is about 3 nm.

It is now of interest to see how this size changes in the semicrystalline sample.

The activation diagram for the semicrystalline sample (Fig. 13) shows a very complicated behavior. Line A corresponds to process A in Fig. 12, and line B corresponds to process B in Fig. 11. Additionally, we find some further local processes (lines C—E), but no typical glass transition curvature. In the thermogram of this sample (Fig. 14; curve 3), we observe an irregularity in the specific heat capacity in the range 280—300 K. With (1) and a half step temperature of about 286 K, we get point 3 in the

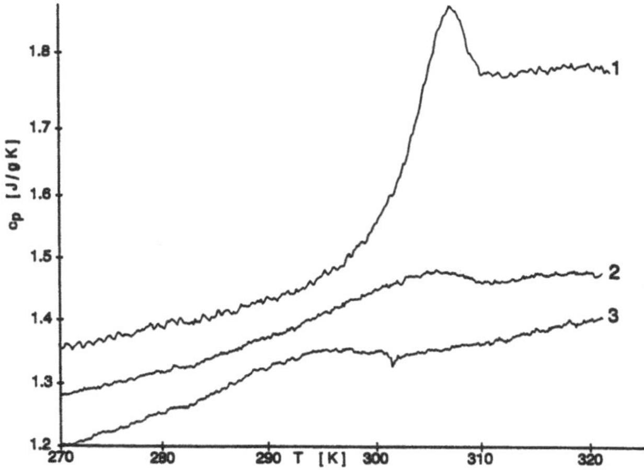

Fig. 14. Thermograms (heating rate 10 K/min) in the glass transition region after cooling the sample at 10 K/min from the melt, and annealing for different time at 310 K. 1 = 0 h; 2 = 3 h; 3 = 13 h. For the sake of clarity the successive curves are vertically shifted. The scale is for sample 3

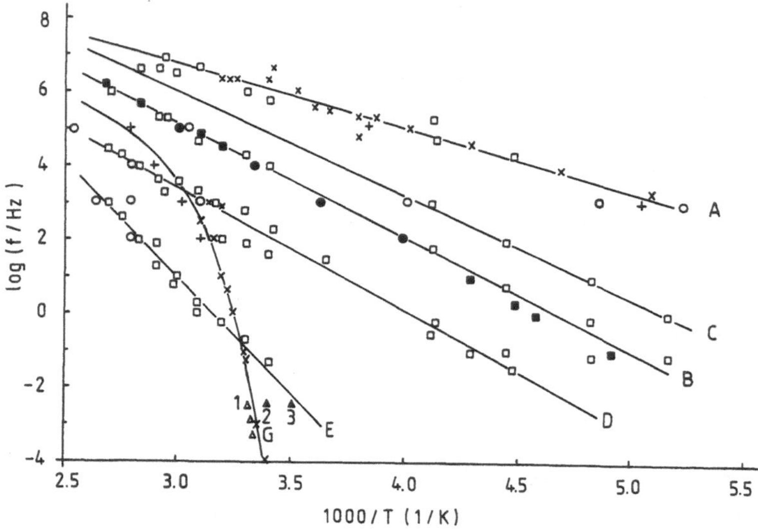

Fig. 15. Activation diagram including all processes observed by dielectric and thermal measurements in the sample with different structure. +; ×; △ = nematic liquid; ■; ● = 100% crystalline; □; ○; ▲ = semicrystalline (T_a = 310 K; t_a = 13 h). The points 1 to 3 demonstrate the structure dependence of the thermal process. 1 = no annealing; 2 = 3 h at 310 K; 3 = 13 h at 310 K

activation diagram (Fig. 15). Assuming that point 3 is part of process E, the transition temperature is lower than the glass transition temperature of the nematic liquid (Fig. 15, point 1).

To show the behavior of the structure dependence of this process, we investigated some samples with a lower degree of crystallinity than sample 3 in Fig. 14. These samples were annealed at 310 K for different times and then measured (Fig. 14 1—0 h; 2—3 h; 3—13 h).

To demonstrate the behavior, all processes are plotted together in Fig. 15. The points (1—3) demonstrate the structure dependence of the thermal process. Compared with the glass transition in semicrystalline polymers, the transition interval is very small for sample 2 and 3 (about 15 K). In semicrystalline PET, we found transition intervals of about 30 K, while the transition intervals of the amorphous samples are nearly the same in both types of materials.

The parameters of all detected processes are presented in Table 1.

Discussion

The study of the relaxation behavior of small amorphous layers makes it possible to obtain information about typical length scales of glass transition. Because there is no great influence of interfacial layers between the crystallites and the liquid (glass-forming) layers, it is useful to investigate low molecular mass semicrystalline materials. 3'3-sulphonyl-bis-[methyl 4-(4-pentoxy-benzoyloxy)-benzoat] forms such semicrystalline structures, as shown in Fig. 5. Up to now, we have had no detailed information about the arrangement and the dimensions of crystalline and amorpahous parts in this material, which is why we cannot perform quantitative calculations. But there are some qualitative results:

i) the relaxation behavior of the nematic liquid is typical for a glass-forming liquid. One local (Arrhenius-like) and one cooperative glass process appear;
ii) in the crystalline and the semicrystalline states, we find only Arrhenius-like relaxation processes;
iii) the thermograms of the semicrystalline sample show the indication of a process which is related to the slowest detected dielectric process in this sample;
iv) the preexponential factors for the processes A to D in Fig. 15 are in the range 10^{12} ... 10^{14} Hz, which is normal for secondary relaxation processes. In contrast, the preexponential factor of process E is about 10^{20} Hz. This cannot be discussed by a simple two-sided model with temperature-independent activation energy.

The behavior of relaxation process E shows no features which can be properly evaluated in terms of Eqs. (3)—(6). Therefore, we cannot draw precise conclusions for the size of cooperatively rearranging regions. But, if we assume that process E is a residuum of the glass transition caused by a strong spatial limitation of the glass-forming parts of the semicrystalline structure, the tentative application of Eqs. (3), (4) yields the qualitative result that the characteristic size is much smaller than in the nematic liquid.

In consequence, it should be possible to investigate the transition from a slightly hindered glass transition, like in semicrystalline PET, to a state where a glass transition is not possible due to the spatial limitation. To control the size of the amorphous regions, we will investigate a series of different liquid crystalline materials, especially with different lengths of the stiff part of the molecules. Further investigations are necessary to get detailed information about structure and relaxation behavior of these interesting materials.

Acknowledgement

The authors wish to express their thanks to Prof. Dr. W. Pechold, Doz. Dr. E. Donth, and B. Götschmann for helpful discussions, and to the Deutsche Forschungsgemeinschaft (SFB 239) for financial support.

References

1. Tammann G (1933) Der Glaszustand. Leopold Voss, Leipzig
2. Angell CA, Sare JM (1978) J Phys Chem 82:2622
3. Moynihan CT, Sasabe H, Tucker J (1976) Proc Int Symp Molten Salts (Ed by IP Pemster et al.), Elektrochem Soc Princetown
4. Shen MC, Eisenberg A (1970) Rubber Chem Technol 43:95, 156
5. Suga H, Seki S (1980) Faraday Discuss 69:221
6. Donth E (1982) J Non Cryst Solids 53:325
7. Pechhold WR, Stoll B (1982) Polym Bull 7:413
8. Owen AJ, Bonart R (1985) Polymer 26:1034
9. Strom U, Tailor PC (1977) Phys Rev B16:5512
10. Schick C, Donth E (1991) Physica Scripta
11. Schick C, Fabry F, Schnell U, Stoll G, Deutschbein L, Mischok W (1988) Acta Polymerica 39:705
12. Schick C, Wigger J, Mischok W (1990) Acta Polymerica 41:137
13. Dehne H, Roger A, Demus D, Diele S, Kresse H, Pelzl G, Wedler W, Weissflog W (1989) Liquid Crystals 6:47
14. Baumeister U, Hartung H, Roger A, Dehne H (1991) Molec Crystals Liq Crystals (to be published)
15. Cheng SZD, Wunderlich B (1987) Macromolecules 20:1630
16. Stoll B (1975) Archiv techn Messen 474:129; Heinrich W (1987) Thesis Ulm
17. Binder B (1989) Thesis Ulm
18. Yalof SA, Hedvig P (1976) Thermochemica Acta 17:301
19. Heinrich W, Stoll B (1988) Progr Colloid Polym Sci 78:37
20. Meischner C, Greiner B, Hauptmann P, Donth E (1986) Acta Polymerica 37:453
21. Pechhold W (1980) Colloid Polym Sci 258:269
22. Schneider K, Schönhals A, Donth E (1981) Acta Polymerica 32:471

Authors' address:

Doz. Dr. C. Schick
Hochschule Güstrow
Institut für Physik
Goldenberger Str. 12
O-2600 Güstrow, FRG

Discussion

HAVRANEK:
What was your measure of crystallization degree?

SCHICK:
The crystallization degree was determined from the heat of fusion by $\alpha = \Delta H/\Delta H_c$, where ΔH is the heat of fusion for the sample under investigation and ΔH_c is the heat of fusion of the 100% crystalline material. In the case of low molecular mass liquid crystal it is possible to measure ΔH_c directly. Up to now, we are not able to get any further information about the structure of the semicrystalline sample, e.g., from x-ray diffraction or electron microscopy.

HAVRANEK:
What is the distribution of the amorphous phase?

SCHICK:
We have no detailed information about it. We think that the distance between growing fronts becomes so small

during crystallization that it is not possible to orientate the stiff molecules in the direction necessary to incorporate the molecule in the crystal. If this is the case, the distance between the crystallites (thickness of the amorphous phase) is controlled by the molecular dimensions and the distribution should not be so broad. Then it should be possible to vary this distance by a variation of the stiff part of the molecules, which we will try in future.

WENDORFF:
Why did you not take blockcopolymers as model systems, since, with them, you can control the morphology by chemical means?

SCHICK:
Donth*) and coworkers tried to do this, and they tried to enclose a glass-forming liquid (glycerol) in porous glasses (diameter of the porous material, some nanometers). In both cases, the problem is that there is an interface between chemically different compounds. Such an interface is connected with a more or less extensive interfacial layer. Estimating the dimension of the interfacial layers and its relaxation behavior is very complicated. Yet it was not possible to get information about glass transitions length scales. Our idea was to use heterogeneous materials which only contain interfaces between phases of the same substance, e.g., semicrystalline structures [10]. In such materials it is possible to estimate the dimension of the liquid regions, as well as the dimensions of the interfacial layers [10, 11].

*) Donth E (1984) Acta Polymerica 35:120

Progress in Colloid & Polymer Science Progr Colloid Polym Sci 85:157—162 (1991)

Laser-induced electrical conductivity in thin layers of poly(bisethylthio-acetylene)

H.-K. Roth[1]), R. Baumann[1,2]), J. Bargon[2]) and M. Schrödner[1])

[1]) Leipzig University of Technology, Department of Natural Sciences, Leipzig, FRG
[2]) University of Bonn, Institute of Physical Chemistry, Bonn, FRG

Abstract: The paper reports on studies of a newly developed material which opens up new ways for the combined use of laser and computer techniques for advanced technological application. Poly(bis-ethylthio-acetylene) (PETAC) is easily soluble in halogenated hydrocarbons and aromatic solvents. Thin films of PETAC can be formed on various substrates such as glass, silicon, ceramics, metal or polymer. The layers are well insulating ($\sigma = 10^{-14}$ Scm^{-1}). Electrically conducting paths or patterns with 10^{-1} $Scm^{-1} < \sigma < 10^2$ Scm^{-1} were directly generated by laser irradiation of PETAC layers with wavelengths of the visible and of the ultraviolet regions. The highest conductivity of laser modified PETAC was obtained by application of an Ar^+-laser operating at λ = 488 nm. The structurization of the materials is also possible by ablation. Sharp patterns were obtained by XeF-excimer laser emitting at λ = 351 nm.

Key words: Conducting polymer; poly(bis-ethylthio-acetylene) (PETAC); laser recording; electron spin resonance (ESR)

Introduction

New developments in technology require, for many purposes, organic polymers with the following combination of properties:

- The polymers should be well soluble in normal organic solvents.
- They should be able to form thin polymer layers on various substrates.
- The polymers have to be stable, even without protection against oxygen or other corrosive gases, against moisture and sunlight.
- By means of intense laser light the polymer should be quickly and locally modifiable under significant changes of solubility, conductivity, and optical properties.

In previous papers, we informed about the development of a suitable basic polymer, poly(bis-alkylthio-acetylenes) and about some conditions for laser-induced conversion of the insulating and well-soluble initial polymer into a highly conductive and completely insoluble material by the application of the radiation of an Argon ion laser [1—4]. Another paper reported on first modification experiments on this polymer by means of excimer laser [5].

The present work is devoted to a more detailed discussion on the conversion process in poly(bis-*ethyl*thio-acetylene) (PETAC), and its material properties, with the aim to get more information on wavelength dependence of the modification process and to find a way to a conversion product with other morphology and better mechanical and electrical properties.

Some properties of the initial polymer

The poly(bis-ethylthio-acetylene) (PETAC) is easily soluble in halogenated hydrocarbons and aromatic solvents. From these solutions thin films can be obtained on glass, silicon, ceramics, metal or polymer surfaces by spin-coating.

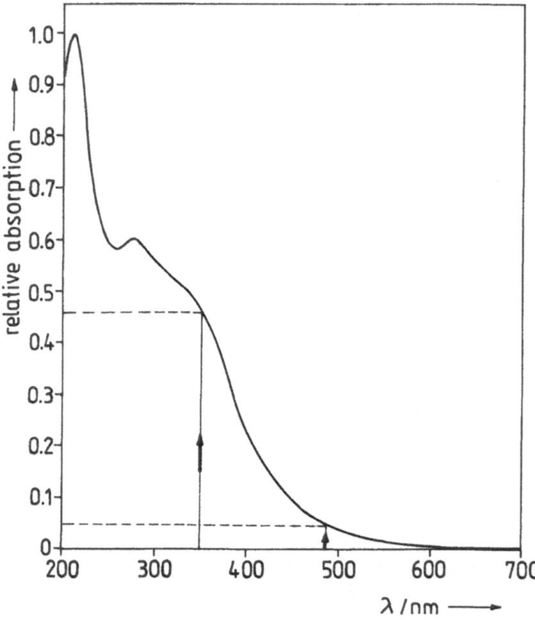

Fig. 1. UV-VIS spectrum of PETAC dissolved in $CHCl_3$. The arrows indicate the applied laser wavelenghts

By means of solid-state ^{13}C-NMR studies, we know that the main chains of PETAC consist of sp^2-hybridized carbon atoms, and that the 1,2 bis-ethylthio-acetylene units build up a nonplanar polymer backbone [3]. In Fig. 1 the UV-VIS spectrum shows a maximum near 215 nm and a broad absorption of low intensity up to 600 nm. The UV-VIS spectrum also indicates a twisted polymer backbone and no extended conjugated π-electron system. This is one reason for the insulating properties of the unmodified PETAC.

From this UV-VIS spectrum, it is evident that the initial polymer can be modified by both ultraviolet and visible light. In the wavelength range of the Argon ions laser (near 488 nm) the absorption coefficient is lower by one order of magnitude than that for UV-lasers.

For that reason the conversion of PETAC by 488 nm radiation requires the application of a well-focused laser beam for recording the conducting paths. An advantage of the relatively low absorption coefficient at λ = 488 nm lies in the possibility to convert also layers of 7- to 10-μm-thickness up to the bottom. The UV-VIS spectrum of PETAC offers a very effective modification or ablation by UV-laser, and especially, this is the case if the layers have a thickness of less than 2 μm.

The thermal behavior of PETAC was studied in the temperature range from 30 °C to 600 °C. All PETAC samples not treated by laser have a thermal stability lower or comparable with that of most of the usual polymers [2]. DTA investigations show the beginning of decompositions at about 160 °C and a decomposition maximum at about 300 °C, [6]. Because of the comparably low stability of the C-S bond to the carbon back bone of the polymer, it is possible to split off the ethylthio groups, which leads to the formation of unsaturated structures with an extended π-system. The conversion is exothermic. Diethyl-disulfides, ethyl-mercaptans and ethyl-disulfides are formed mainly as splitting products. The residue is a black solid, insoluble in all solvents. Although the pyrolysis at temperatures above 300 °C results in conversion products mainly composed of sp^2-hybridized carbon and sulfur, as proved by means of solid-state ^{13}C-NMR spectroscopy [3, 6], the *products of pyrolysis* are still insulators.

Recording of conducting paths in PETAC by Ar^+-laser

For this purpose, poly(bis-ethylthio-acetylene) dissolved in $CHCl_3$ was cast on polyimide or glass substrate. The thickness of the dry PETAC film usually was in the range of 1.5—7 μm. The conversion was performed by means of a medium-power cw-Argon ion laser at λ = 488 nm. The exposure of polymer samples was made line-by-line by means of a computer-controlled positioning system. A defined threshold of laser intensity is necessary to convert the polymeric insulator into an electrically conductive material. At laser intensities higher than 15 kW cm^{-2} the scan velocity can be about up to 10 ms^{-1} [2]. The attainable conductivity depends both on the scan velocity and on the laser intensity in a nonlinear manner, and additionally, on the reflexion and the heat conductivity of the substrate. If the applied laser intensity is too low, the irradiated PETAC remains a well-insulating material. When the laser intensity used approaches the threshold intensity a jump-like increase of conductivity is observed. The start of this conversion is accompanied by the generation of free radicals.

ESR measurements represented in Fig. 2 show that with increasing laser intensity the spin concentration increases from $4 \cdot 10^{14}$ spins g^{-1} (unexposed polymer) to $5 \cdot 10^{17}$ spins g^{-1}. Upon a further increase of laser intensity, the number of para-

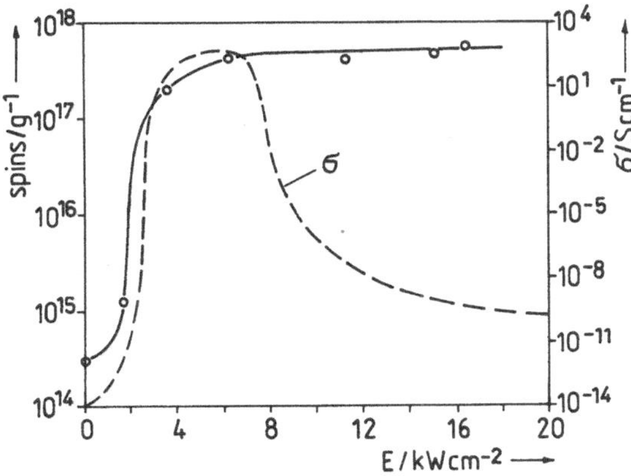

Fig. 2. Electron-spin concentration in laser-converted PETAC as measured by ESR at room temperature after irradiation by Ar^+ laser (full line) and a qualitative illustration of a possible intensity dependence of dc conductivity

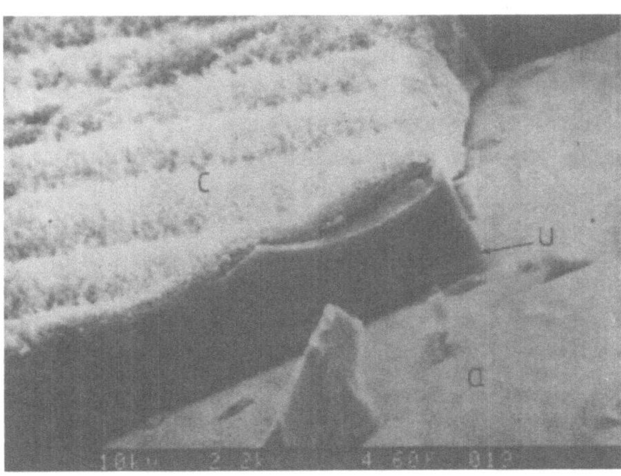

Fig. 3. Electron micrograph (SEM record) of PETAC partly converted by excimer laser in presence of air. a: ablated area; c:converted material; u: unconverted PETAC

magnetic centers is approximately constant [2]. This does not apply to the electrical conductivity. The latter depends more strongly on the irradiation conditions. Whereas the radical concentration during conversion increases by only 3 orders of magnitude, the conductivity changes by 16 orders. However, under some experimental conditions, if laser intensity is too high, electrical conductivity again decreases. A possible intensity dependence is schematically illustrated in Fig. 2. This effect is still under investigation and will be published in another paper.

Generation of conducting patterns in PETAC by XeF-excimer laser

The preparation of polymer layers was made in the same way as in the case of laser recording by argon ion laser. For example, sheets of glass were plated with thin layers of PETAC by spin-coating using a solution of 9 wt% PETAC in $CHCl_3$. The only exception was that thinner layers (of a thickness of less than 2 µm) were preferred because of the low penetration depth of UV light into the polymer due to the high absorption coefficient. The different absorption (seen in Fig. 1) effects that, according to the law of Lambert and Beer, a 0.5-µm layer of PETAC absorbs more energy at λ = 351 nm than a 5-µm layer at λ = 488 nm. In the electron microscopic picture of Fig. 3 it is well seen that from

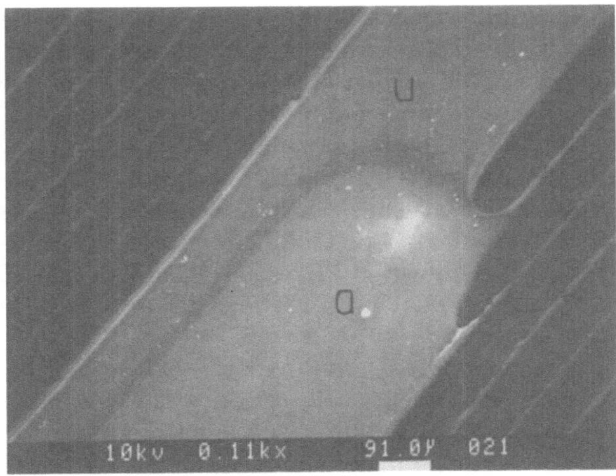

Fig. 4. Conducting pattern (dark areas) in two laser wavelength processed PETAC. First pattern by ablation using a 351 nm XeF excimer laser (area a). Afterwards the PETAC was partly converted line-by-line, into a conducting form using a 488 nm Ar^+-laser. Area u: unconverted PETAC

a PETAC layer of about 7-µm thickness only a very thin layer was converted by the excimer laser radiation.

In the case of material modification by excimer laser, stainless steel masks were used for the generation of conducting patterns. In some experiments the patterning was made by ablation of insulating PETAC at λ = 351 nm or 248 nm. In an additional step the insulating PETAC pattern was converted, line-by-line, into a conducting material by Ar^+-laser. Figure 4 shows such a pattern. For the XeF-excimer laser radiation (λ = 351 nm) applied, the energy threshold is about 80 mJ cm^{-2} to remove a polymer layer of 1-μm thickness [5].

The energy necessary for modification or ablation of PETAC layers is lower than that for most of the other polymers [7–9], which allows a high speed in material modification or processing. This is due to the ethylthio groups. The C-S bond to the polyacetylene back bone has a low binding energy. It is, compared with C-C and C-H bonds, distinctly weaker. Additionally, large ethylthio groups weaken the double bonds between carbon atoms which facilitate laser-induced chemical conversion.

Properties of laser-converted PETAC

The laser-converted PETAC is insoluble. As discussed in a previous paper [2], the electrical conductivity of conducting paths recorded by argon ion laser is in the range from 1 S cm^{-1} to 200 S cm^{-1} depending on layer preparation, on the applied substrate and on laser irradiation conditions. For example, on a glass substrate scan velocity of 5 mm/s and a laser intensity of 15 kW cm^{-2} gives a conductivity in the converted PETAC layer of about 100 S cm^{-1}, as well as a conversion on a polyimide substrate at 2 m/s at 17 kW cm^{-2}.

In first experiments, in the case of 351 nm XeF excimer laser irradiation at a fluence of 3 J cm^{-2}, a conductivity of $1.5 \cdot 10^{-1}$ S cm^{-1} was found.

The conducting material is long-time stable, even in the presence of moisture and in a corrosive atmosphere. The charge carrier mobility of the laser converted PETAC is higher than in most of the other polymers and also higher than in amorphous silicon. It depends on sample preparation and irradiation conditions. According to a first estimation from Hall coefficient measurements, it lies between 10^{-1} and 10 cm^2 V^{-1} s^{-1} at room temperature and increases with lowering the temperature [2]. Due to the opposite temperature dependence of the charge carrier concentration and of the carrier mobility, the laser converted PETAC shows an extremely low temperature coefficient of the electrical resistance ($3 \cdot 10^{-4}$ K^{-1} < α < $2 \cdot 10^{-3}$ K^{-1}). The conducting PETAC has a very low frequency dependence of conductivity. It is a *p*-type semiconductor.

The color and the morphology of the converted PETAC depend on the irradiation wavelength and on the degree of modification. The unconverted insulating PETAC layers have a yellow or yellow-brown color. The conversion of the insulator into a conductive material of medium conductivity by excimer laser irradiation is not accompanied by a significant change of color; it becomes only a little darker, whereas the highly conductive material generated by Argon ion laser is a blue-black product. The electron micrograph in Fig. 5 shows that the morphology of the Ar^+-laser converted paths differ very strongly from the unconverted ones. The initial PETAC layers are almost homogeneous and smooth, whereas the conversion products have a porous structure. Probably, the porous structure is due to the vehement exit of gaseous reaction products. Using the excimer laser radiation for conversion, we found a velvet-like structure of the conducting material as seen in Fig. 3 [5].

Figure 6 shows the sulfur distribution across the unconverted and converted PETAC tracks. (The conducting paths were here recorded by Ar^+-laser.) The spatially resolved elementary analysis made by an x-ray microprobe technique

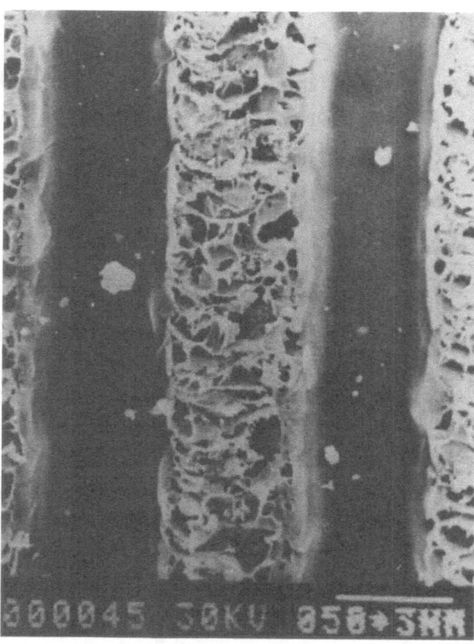

Fig. 5. Electron micrograph of unconverted (u) and laser-converted (c) paths (converted by Ar^+ ion laser in presence of air)

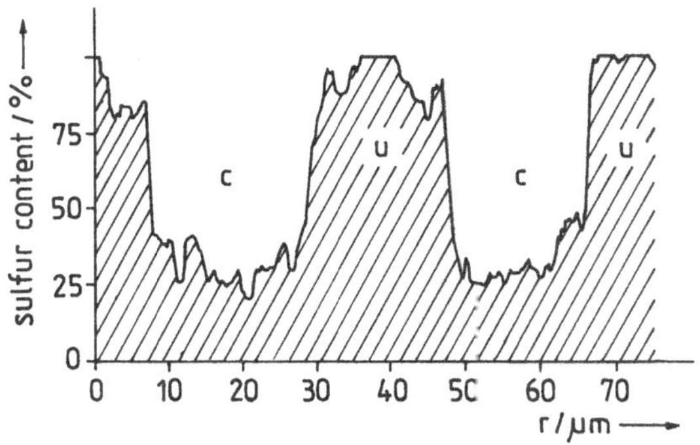

Fig. 6. Sulfur distribution in PETAC across the paths after localized laser-induced conversion at λ = 488 nm and room temperature in the presence of air, measured by x-ray fluorescence microprobe. u: unconverted PETAC layer; c: converted areas

shows that there is still a notable amount of sulfur, also in the converted regions. The drastic difference in the sulfur concentrations between the converted and unconverted areas (within 1—2 µm) is due to the existence of a defined threshold energy for the conversion. This observation agrees very well with the results of ESR measurements.

The ESR spectra of the initial PETAC and of the laser-converted products differ very strongly in signal intensity, but all consist of a single line or of a superposition of single lines. The insulating (unconverted) polymer has a g-factor of 2.0056. It decreases by laser-induced conversion to g = 2.0039, which indicates that, as a result of polymer modification, the unpaired electrons still interact with the sulfur, but less so than in the initial PETAC [2]. However, the g-factor of the conducting PETAC modification also differs significantly from that of radicals in pure hydrocarbons or in other polymers in which the unpaired electrons do not interact with heavy atoms like sulfur and exhibit a g-factor of about 2.0023. From the measurements it can be concluded that the laser-induced conversion of PETAC is no graphitization process, but it is sure that the fast reaction leads to a product which includes sulfur bonds in positions touched by the unpaired electrons which contribute to the electrical conductivity.

Taking into account reaction models developed in connection with the study of the thermal degradation of PETAC [6] it can be assumed that by means of Ar^+-laser radiation a photopyro-conversion takes place, which leads to the formation of unsaturated structures with an extended π-system mainly composed of sp^2 hybridized carbon atoms which are partly chemically linked with sulfur.

Applicability in electronics

The high conductivity, the long-time stability of the material, and the low temperature coefficient of the resistance make the laser modifiable PETAC especially suitable for application in information electronics. Using a computer-controlled laser direct imaging system, one is able to manufacture printed circuit boards, with conducting micropaths and directly recorded passive devices such as resistors, capacitors, and coils directly in the circuit board, in order to realize high integration.

For example, we obtained capacitors by converting the polymer in comb-like structures into the conducting form. The width of a single track as well as the distance between two tracks was less than 20 µm. By use of unconverted PETAC as dielectricum, we got capacitors with a specific capacity of more than 20 pF per cm^2. In a comparable procedure resistors can be obtained by writing the tracks in a meandering form and by tuning the irradiation conditions to result in lower conductivity of the laser-converted PETAC. Other elements, especially surface-mountable devices, can be connected by conductive adhesives, but also by soldering or bonding. Laboratory models of such circuits have been successfully tested. For example, a small multivibrator wiring made of laser-recorded conducting paths and resistors, completed by SMD components, have been working stable for more than 6 months.

Acknowledgement

The studies using the excimer laser have been financially supported by the "Fonds der Chemischen Industrie". We thank Dr. A. M. Richter and Prof. Dr. E. Fanghänel from the Merseburg Technical University for the initial PETAC, for valuable discussions, and for providing us with the manuscript of their paper [6].

References

1. Richter AM, Richter JM, Beye N, Fanghänel E (1987) J Prakt Chem 230:811
2. Roth HK, Gruber H, Fanghänel E, Richter AM, Hörig W (1990) Synth Met 37:151

3. Hempel G, Richter AM, Fanghänel E, Schneider H (1990) Acta Polymerica 41:522
4. Roth HK, Baumann R, Schrödner M, Gruber H (1991) Synth Met 41:141
5. Baumann R, Bargon J, Roth HK (1990) Proceedings of the SPIE Conference, USA
6. Richter AM, Laube U, Morisak F, Fanghänel E, Hempel G, Goldenberg S, Scheler G, Acta Polymeica (in press)
7. Decker C (1987) J Polym Sci C Polym Lett 25:5
8. Decker C (1989) Macromol Chem Macromol Symp 24:253
9. Davenas J, Boiteux D, Adem EH, Sillion B (1990) Synth Met 35:195

Authors' address:
H.-K. Roth
Leipzig University of Technology
Department of Natural Sciences
P.O. Box 66
7030 Leipzig, FRG

Discussion

KRYSZEWSKI:

It is not easy to determine the charge carrier mobility from Hall measurements on polymers. How reliable are your estimations?

ROTH:

We know that it is very difficult to have our laser-modified polymer samples correctly fulfil all conditions for the determination of charge carrier mobility by means of Hall coefficient measurements. What we have done is to measure our polymer samples under the same instrumental conditions as the ordinary anorganic semiconductors, and to use the same formulas for the estimation of mobility. For this reason, we should not overrate the estimated charge carrier mobilities. Nevertheless, we are reasonably sure that the observed temperature-dependence of the charge carriers is correct. Further, it can be assumed that the carrier mobility really lies between 10^{-1} and 10 $cm^2\ V^{-1}\ s^{-1}$, but it is impossible to say whether the estimated maximum values are realistic. This would require a better realization of the conditions for reliable Hall coefficient measurements, but this we cannot do. The applied Hall measurements are valuable, in the first place, for comparison of various samples. They allow to state that the charge carrier mobility in the laser converted PETAC is higher than in most of the other polymers, and also higher than in amorphous silicon which opens up applicabilities in electronics.

WEICHART:

Is it possible to achieve a temperature coefficient of zero by laser irradiation or doping of the materials you used?

SCHRADER:

Please make some comments on the physical reasons for the low dependence on temperature.

ROTH:

The observed temperature-dependence of the electrical conductivity in laser-converted PETAC is extremely low, but the coefficient is not zero. This behavior is more similar to constant than to ordinary metals or semiconductors, and this makes the material valuable for several purposes. Probably, the physical reason is a kind of compensation effect. A typical property of semiconductors is that the number of charge carriers increases with increasing temperature. As is known from Hall measurements on laser-modified PETAC, the charge carrier mobility depends on temperature in an opposite direction: It can be assumed that this is the reason for the negligible temperature dependence of the electrical conductivity.

WEICHART:

The extremely low temperature coefficient of the electrical resistance probably makes possible some interesting applicabilities in the field of sensorics, for example, for the seperate detection of pressure fluctuations which are caused by or are acompainied with temperature variations. Have you studied the pressure dependence of the resistance of modified PETAC or other properties of your polymer which are essential for the application as a sensor?

ROTH:

No, we have not yet conducted tests or studies in this direction, but it is, without doubt, an interesting subject. In fact, we expect a dependence of the electical resistance of laser-modified PETAC on pressure. However, the utilization of this effect in a sensor will not be very easy, because conducting PETAC is a brittle materal of a porous structure and requires a mechanical protection, for example, by means of other polymers.

Phase transitions in single-crystalline polymerizing diacetylenes as seen by dielectric measurements

M. Orczyk and J. Sworakowski

Institute of Organic and Physical Chemistry, Technical University of Wrocław, Wrocław, Poland

Abstract: This paper reports on results of measurements of low-frequency electric permittivity of mixed single crystals of solid-state polymerizable diacetylenes pTS and pFBS. The measurements were carried out on monomers and fully polymerized samples over the entire composition range, in the temperature range 80—330 K. The phase transitions in pure pTS and poly-pTS are visible on the $\varepsilon(T)$ dependencies, whereas similar dependencies measured in pFBS and poly-pFBS are completely featureless, confirming absence of any phase transitions in that material within the accessible temperatures. In mixed crystals containing increasing amounts of pFBS, the transitions shift towards low temperatures, and their signatures become smeared and less pronounced. Such behavior can be explained by assuming that dipole-dipole interactions are responsible for the phase transitions.

Key words: Polydiacetylene; dielectric properties; phase transition; polymerization

1. Introduction

Measurements carried out on poly(diacetylenes) offer a unique opportunity to study properties of highly organized (single-crystalline) polymers (e.g., [1, 2] and references therein). Moreover, one can also follow an evolution of those properties during the solid-state polymerization which, in some cases, is a topochemical process [3], proceeding in a controlled way from monomer (a typical, nearly isotropic, molecular crystal) to polymer (highly anisotropic, quasi-one-dimensional material, which may be regarded as covalent along the direction of the polymer chains)

$$n\ \mathrm{R{-}C{\equiv}C{-}C{\equiv}C{-}R} \rightarrow (=\overset{\mathrm{R}}{\mathrm{C}}-\mathrm{C}{\equiv}\mathrm{C}-\underset{\mathrm{R}}{\mathrm{C}}=)_n\ .$$

Among the solid-state polymerizable diacetylenes, much attention has been paid to pTS [bis(*p*-toluenesulfonate) of hexadiynediole; R = $-CH_2-OSO_2-\Phi-CH_3$], Since the early papers by Wegner [4, 5], kinetics and energetics of polymerization, as well as properties of monomer and polymer pTS have been extensively studied (e.g., [1, 2] and references therein). Another interesting diacetylene is pFBS [bis(*p*-fluorobenzenesulfonate) of hexadiynediole; R = $-CH_2-OSO_2-\Phi$-F], which has practically identical molecular conformation and is isomorphous with pTS [6—9].

In both monomer and polymer of pTS, phase transitions were found: the monomer undergoes a first-order transition at ca. 160 K, followed by a second-order one at ca. 200 K, whereas only one transition of second order was found in poly-pTS at ca. 190 K [10—12]. These transitions could be conveniently observed by employing dielectric measurements [13, 14]. It is interesting to note that neither monomer nor polymer of pFBS exhibit any phase transition down to helium temperatures [15, 16].

This contribution reports on results of measurements of the evolution of electric permittivity carried out in pTS, pFBS, and a series of their mixed crystals, as a function of temperature. The paper supplements our previous publications that describe results of measurements of dielectric properties of single crystals of polymerizing pTS [13, 14] and pFBS [16].

2. Experimental

The diacetylenes were synthesized by Prof. M. Bertault. The material was then purified and the crystals grown from acetone solutions according to the procedure described in [17]. The composition of mixed crystals was determined from NMR and IR measurements, as described elsewhere [18]. The samples employed in the measurements were cut in the desired directions with a wire saw and then polished on tissues soaked with ethyl acetate. The thicknesses of the samples were ca. 0.5 mm; their electroded area amounted to about 20 mm^2.

The electric permittivities (ε) were calculated from capacitances measured at 1 kHz in the temperature range 80—333 K, in the dark, and under ambient nitrogen pressure. The samples were polymerized in situ at 333 K, the polymer amount being determined directly from the time-conversion dependencies [17, 19, 20].

3. Results

It was found in our earlier measurements that the polymerization substantially modifies electric permittivities of the crystals measured in the direction parallel to that of growing polymer chains (crystallographic **b** direction; ε_b or ε_2). As seen in Fig. 1, the transitions are clearly visible on the $\varepsilon(T)$ curves measured in pure pTS, both in monomer and in fully polymerized samples (curves *a* in Figs. 1a and 1b). The high-temperature phase transition in monomer and monomer-rich samples, as well as the phase transition in polymer manifest themselves as maxima on the $\varepsilon(T)$ curves, whereas the low-temperature transition is visible as a small hump on the low-temperature shoulders of the curves. The other two principal components of the ε tensor were found to vary to a much lesser degree: independently of the conversion to polymer, ε_3 is a weak function of temperature and ε_1 remains practically constant over the entire temperature range under investigations [13, 14].

With increasing concentration of pFBS, the signatures of all transitions shift towards the low temperatures, the $\varepsilon(T)$ curves becoming smeared. As is seen in Fig. 1 (curves *b* through *e*), the trace of the low-temperature transition in monomer samples is lost for the mole fraction of pFBS (y) exceeding 0.05, whereas the high-temperature transition can be followed until $y \approx 0.15$. In fully polymerized samples, the transition is much better marked, the maxima being observed even for $y \approx 0.2$. In samples richer in pFBS, the transition apparently shifts below the temperature region covered in our experiments, and for $0.2 < y < 0.5$ one can observe only the high-temperature branches of the bell-shaped anomalies associated with the transition (cf. curves *g*, *h* and *i* in Fig. 1b).

Contrary to pTS, permittivities of the monomer and polymer pFBS crystals are practically independent of temperature over the entire range under study, exhibiting no evidence of a phase transition (curves *k* in Figs. 1a and 1b).

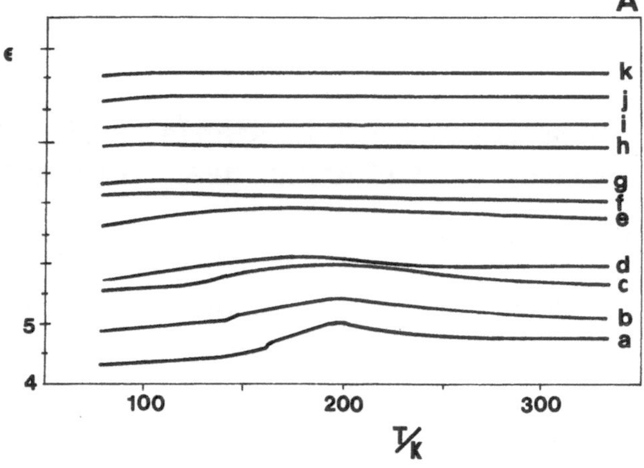

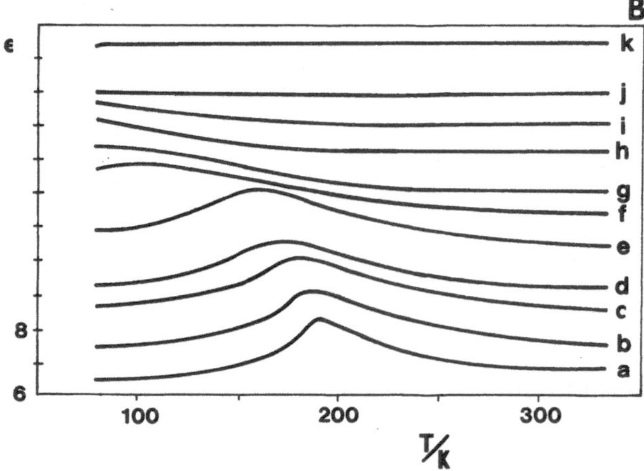

Fig. 1. Temperature dependencies of ε_b in pTS — pFBS diacetylene mixed crystals of various compositions (expressed in mole % of pFBS): a = 0 (pure pTS); b = 3; c = 5; d = 10; e = 15; f = 171 g = 19; h = 26; j = 55, and k = 100 (pure pFBS). A) crystals of monomers; B) fully polymerized samples. For the sake of clarity, curves *a* through *k* on both figures are vertically shifted. (Note different ε scales in A) and B).)

4. Discussion

It was demonstrated by several authors that the phase transitions in pTS are associated mainly with motions of (highly polar) side groups, the central diacetylenic fragment of the molecule playing a negligible rôle. In the low-temperature phase, the tosyl groups are ordered, and assume (in a defined manner) one of two possible orientations, whereas in the high-temperature phase, the positions of the side groups are averaged. As was shown in our earlier papers [13, 14], the transition in polymer (and the high-temperature transition in monomer) can be described as an antiferro-to-paraelectric one, the polarizations of the sublattices being the order parameter.

In spite of close structural similarities of pTS and pFBS [6—9], no phase transition has been observed in both monomer and polymer of the latter compound, probably due to differences in the dipole-dipole interaction energy [16]. The electric moment of the tosyl group amounts to ca. 4 Debye units [21]; replacing the methyl group by the highly electronegative fluorine atom results in a decrease of this value by ca. 1.8 D [22]. Such a significant difference in the dipole moments is likely to influence their mutual interactions responsible for the presence (or absence) of the phase transitions. The reasoning presented above can be verified experimentally, as the mixed pTS-pFBS system offers the possibility of a continuous modification of average energy of dipole-dipole interactions due to unlimited miscibility of pTS and pFBS in the solid phase.

The results shown in Fig. 1 can be qualitatively explained within a simple model [23, 24]. Here, we shall limit ourselves to the transition taking place in fully polymerized samples. Although there exist results indicating its displacive character [25], the phase transition will be assumed to be of the order-disorder type for the sake of simplicity. The crystal can be represented by a network of dipoles, which are ordered in the low-temperature phase, and disordered above the transition. As is shown elsewhere [23, 24], the model predicts that the temperature of the phase transition should follow the dependence

$$kT_{\text{trans}} = \gamma[(1-y)\mu_H + y\mu_G]^2 , \qquad (1)$$

where μ_H and μ_G stand for the dipole moments of the host (p-tosyl) and guest (p-fluorobenzenesulfonyl) groups, y is the mole fraction of pFBS molecules, and γ is a constant independent of the composition of mixed crystals. Hence the shift of the phase transition with respect to that of pure pTS (T^0_{trans}) should amount to

$$\frac{T_{\text{trans}} - T^0_{\text{trans}}}{T^0_{\text{trans}}} = y^2(1-m)^2 - 2y(1-m) , \qquad (2)$$

where $m = \mu_H/\mu_G$.

As becomes evident from Fig. 2, the predicted behavior is indeed met in our experiments. The agreement is almost quantitative for dilute solutions ($y \leqslant 0.15$). The experimentally observed $T_{\text{trans}}(y)$ dependence is best fitted by Eq. (1) for $m \approx 0.4$, quite close to the expected value of that parameter.

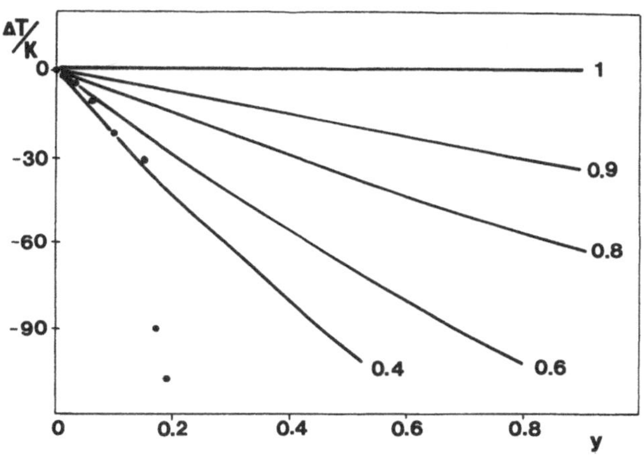

Fig. 2. The shift of the transition temperature in mixed crystals of fully polymerized pTS — pFBS samples. The experimental results are marked as points, the full lines were calculated from Eq. (2) for several values of the parameter $m = \mu_G/\mu_H$

The rapid decrease of T_{trans} in crystals containing more than 15 mole % of pFBS, which cannot be explained within the simple model employed in this paper, might be an indication of aggregation of pFBS molecules at higher concentrations of the guest molecules. However, further experiments are necessary to clarify this point.

Acknowledgements

Thanks are due to Prof. M. Bertault for the gift of pTS and pFBS. The research reported in this paper was supported by the Polish Academy of Sciences within the Programme CPBP 01.14.

References

1. Cantow HJ (ed) (1984) Polydiacetylenes. Springer, Berlin
2. Bloor D, Chance RR (eds) (1985) Polydiacetylenes: synthesis, structure and electronic properties. Nijhoff, The Hague
3. Wegner G (1980) Faraday Disc 68:494–508
4. Wegner G (1971) Makromol Chem 145:85–94
5. Wegner G (1972) Makromol Chem 154:35–48
6. Kobelt D, Paulus EF (1974) Acta Cryst B30:232–234
7. Enkelmann V (1977) Acta Cryst B33:2842–2846
8. Aimé JP, Lefebvre J, Bertault M, Schott M, Williams JO (1982) J Phys (Paris) 43:307–322
9. Aimé JP, Schott M, Bertault M, Toupet L (1988) Acta Cryst B44:617–624
10. Enkelmann V, Wegner G (1977) Makromol Chem 178:635–638
11. Robin P, Pouget JP, Comes R, Moradpour A (1980) J Phys (Paris) 41:415–421
12. Bertault M, Collet A, Schott M, J Phys (Paris) Lettres 42:L131–L133
13. Nowak R, Sworakowski J, Kuchta B, Bertault M, Schott M, Jakubas R, Kołodziej HA (1986) Chem Phys 104:467–476
14. Orczyk M (1990) Chem Phys 142:485–493
15. Chance RR, Yee KC, Baughman RH, Eckhardt H, Eckhardt CJ (1980) J Polym Sci Polym Phys Ed 18:1651–1661
16. Orczyk M, Sworakowski J, Bertault M, Faria RM (1990) Synth Metals 35:77–81
17. Bertault M, Schott M, Brienne M, Collet A (1984) Chem Phys 85:481–490
18. Orczyk M, Pater E, Sworakowski J (1991) Makromol Chem, in press
19. Bloor D, Koski L, Stevens GC, Preston FH, Lando DJ (1975) J Mater Sci 10:1678–1688
20. Baughman RH (1978) J Chem Phys 68:3110–3121
21. Eucken A (ed) (1951) Landolt-Börnstein Zahlenwerte und Funktionen. Springer, Berlin, 6th Ed, Vol 1, Pt III, p 446
22. Minkin VI, Osipov OA, Zhdanov YA (1968) Dipolnye momenty v organicheskoi khimiyi. Izd Nauka, Leningrad, Ch 3
23. Orczyk M (1990) Thesis, Technical University of Wrocław
24. Orczyk M, Sworakowski J, to be published
25. Terauchi H, Ueda T, Hatta I (1981) J Phys Soc Japan 50:3472–3475

Received January 3, 1991
accepted February 8, 1991

Authors' address:

Dr. Maciej Orczyk
Prof. dr Juliusz Sworakowski
Institute of Organic and Physical Chemistry
Technical University of Wrocław
Wybrzeże Wyspiańskiego 27
50-370 Wrocław, Poland

Discussion

GERHARD-MULTHAUPT:

Could you imagine to break the symmetry of the antiferroelectric phase by an electric field in a manner similar that of $a - a_p$ transformation in PVDF?

SWORAKOWSKI:

The formal description of the low-temperature phases of pTS and poly-pTS as antiferroelectric ones does not necessarily mean that a polar ordering of dipoles of both sub-lattices is possible under the action of external electric fields. For example, fields necessary to achieve the polar arrangement may exceed the breakdown field. In pTS, no experimental evidence of such an effect has been found for biasing fields exceeding 10^6 V/m.

Author Index

Althausen D 66

Bargon J 157
Baumann R 157
Biedermann H 118
Birshtein TM 38
Bittrich HJ 82
Borisov OV 38
Brehmer L 46
Bubeck C 143

Chudáček I 1

Erokhin V 47

Feigin LA 47
Fleischer G 127
Freidzon YS 125

Gerhard-Multhaupt R 133
Geschke D 60, 127
Gnoth M 148

Hamann C 102
Havránek A 119
Heckner KH 75
Heilmann A 102
Heise B 12
Hergeth WD 82
Holstein P 60

Kapitza H 124
Kilian HG 12
Koch KH 143
Köpp E 59
Kremer F 124
Kryszewski M 91

Mathy A 143
Mercurieva AA 38
Möhwald H 52
Müllen K 143
Müller J 111

Orczyk M 163

Paul E 12
Pechhold W 59
Peskova E 66
Peterson IR 52
Plümer F 76
Poths H 124

Roger A 148
Roth HK 157

Sautter E 59
Schawe J 148
Schick C 148
Schmutzler K 82
Schrader S 143
Schrodi W 12
Schrödner M 157
Shibaev VP 125
Slavinská D 1, 119
Spěváček J 60
Stamm M 55
Steinau UJ 82
Stoll B 148
Sworakowski J 163

Thiele V 60

Uhlig A 75

Vallerien SU 124

Wartewig S 82
Wegner G 143
Weichart J 111
Wendorff JH 126
Wróbel AM 91
Wübbenhorst M 23
Wünsche P 23, 66

Zentel R 124
Zhulina EB 38

Subject Index

activation parameters 148

behavior, thermal 148
blends 55

charge, surface 1
composite 102
—, polymer 85
conducting polymer 157
core-shell structure 85
crystal, liquid 148
crystalline structure 119
crystals, polymer liquid 38
current, photo-induced 66

decomposition, spinodal 66
defect morphology 66
deformation, large 12
depolarization 66
deposition 91
derivates, perylene 143
dielectric properties 163
— relaxations 148
dispersed liquid crystal, polymer- 133
distribution of polarization 23

elastomer 133
electric poling 60
electro-optic layer 133
emulsion polymerization 85
entanglement-network 12
ESR — electron spin resonance 157

film, organosilicon 91
films, Langmuir-Blodgett 47
—, multilayers 23
—, polymer 66, 102
—, protein 47
fluoride, polyvinylidene (PVDF) 23

gas permeation 111
generation, third harmonic 143
glass transition 148
grafted polymer layers 38

harmonic generation, third 143

interdiffusion 55
interfaces 55
interfacial layer 85

Langmuir-Blodgett films 47
laser recording 157
layer, interfacial 85
layers, plasma polymerized 119
—, polymer 127
light modulator, spatial 133
limitations, spatial 148
liquid crystal 148
— —, polymer-dispersed 133
liquid-crystalline polymers 127
low-dimensional systems 38

membrane 111
—, polymer 133
membranes 47
metal particles 102
model, van der Waals-network 12
modulator, spatial light 133
monomolecular polymer layers 38
Monte-Carlo simulation 66
morphology, defect 66
multilayer films 23

network 119
—, entanglement 12
neutron reflectivity 55
NMR, ^{13}C CP/MAS 60
NMR, solid-state ^{1}H 60
nonlinear optical properties 143

optical properties, nonlinear 143
order, orientational 127
ordering, orientational 38
organosilicon film 91
orientation 66
orientational order 127
— ordering 38

parameters, activation 148
particles, metal 102
permeation, gas 111
perylene derivatives 143
phase transition 38, 163
photo-induced current 66
plasma polymer 102
— polymerization 91, 111
— polymerized layers 119
polarization, distribution of 23
poly(bis-ethylthio-acetylene) (PETAC) 157
poly(N-vinylcarbazole) (PVCa) 66, 119
polyamide 11 (PA 11) 60
polydiacetylene 163
polyethylene 12
polymer 85
— composites 85
— conducting 157
— films 66, 102
— layers 127
— —, grafted 38
— —, monomolecular 38
— liquid crystals 38
— membrane 133
—, plasma 102
polymer-dispersed liquid crystal 133
polymerization 163
—, emulsion 85
—, plasma 91, 111
polymerized layers, plasma 119
polymers, liquid-crystalline 127
polystyrene 1
polyvinylidene fluoride (PVDF) 23
profile, surface 119
properties 91
—, dielectric 163
—, nonlinear optical 143
protein films 47
proteins 47

recording, laser 157
reflectivity, neutron 55
—, x-ray 55
relaxations, dielectric 148
roughness 1, 119

silicone 111
solid-state ^{1}H NMR 60
— NMR 127
spatial light modulator 133
— limitations 148
spinodal decomposition 66
structure 91
structure, core-shell 85
—, crystalline 119
surface charge 1
— profile 119
— tension 1
systems, low dimensional 38

tension, surface 1

thermal behavior 148
thermoplast 133
third harmonic generation 143
transition,
glass 148
—, phase 38, 163
van der Waals-network model 12
x-ray reflectivity 55